The Green Phoenix: Reforestation and Recovery in Northeast Ohio

Kyile

First Printing, 2024

TABLE OF CONTENTS

CHAPTER I

HISTORY AND CHALLENGES FACING THE GREAT LAKES FORESTED AREA

The world has seen a general loss of forested environments throughout recent history, generally done for agricultural use or timber production (Foley et al., 2005). The forested environments of Northeast Ohio and surrounding areas of the Great Lakes region have undergone a large amount of change ever since settlement began in the late 18th century. However by the late 19th century, this once agricultural land was abandoned and secondary forests became established after the second half of the century (Fischer et al., 2013). These land use changes and general disturbances has resulted in a general trend of severe deforestation followed by reforestation efforts (Deines et al., 2016). Environments which have not been influenced by any human activity are referred to as primary forests, but forests that have been cut and developed on past agriculture sites are commonly referred to as secondary forests (Flinn et al., 2005).

Beyond this classification there exist many other metrics that distinguish secondary forests from their primary counterparts. Secondary forests are characterized by a shift in overstory and understory species composition. Often due to the large disturbance caused by agricultural land use, forest understory vegetation is significantly lower in secondary forests compared to primary forests (Singleton et al., 2001). The overstory of secondary forests are also marked by a lower species diversity compared to pre-disturbance forests (Schulte et al., 2007). Beyond species composition, secondary forests lack structural complexity such as snags, or standing dead wood and other coarse woody debris (Foster et al., 2003). This lack of woody debris is linked with human activity and results in less carbon storage in forest environments (Harmon et al., 2004). This altered nutrient cycling appears in both total ecosystem nutrient composition and the soil nutrient composition. It has been consistently reported that the conversion of forest land to agricultural land has resulted in a decrease in soil carbon (Murty et al., 2002). This is further compounded by secondary forests sequestering carbon and nitrogen in the plant biomass rather than mineral stores in the soil (Hooker & Compton, 2017). This change in carbon sequestration pools can have influences on the soil microbial community which can have direct influences on the plant species diversity as well (Brockerhoff et al., 2017).

The management of secondary forests commonly falls into two main categories, of reaching sustainable timber production or increasing these secondary forests health and providing ecosystem services (Kern et al., 2017). However, oftentimes the managed land will have goals that fall into both categories which complicates the already unique nature of the secondary forest (Gustafsson et al., 2012). The goals of managing forests is further complicated as managers need to approach management opportunities while considering how their forested area may respond to an ever changing climate (Frelich et al., 2020).

As forest management practices have evolved, a consistent goal of mimicking natural disturbances has become more present. Common natural disturbances include fires, wind, insect or pathogen outbreaks, yet these disturbances rarely cause full stand break down (Franklin et al., 2002). These disturbances instead create small gaps in the canopy which help to facilitate succession development through increased light levels at the forest floor. In addition to increased light levels at the forest floor, these canopy caps caused by disturbances have cascading influences on the forest ecosystem. Canopy gaps have increased temperature as well as more water within the soil (Muscolo et al., 2014) The size of the canopy gap is known to also influence how successful different shade tolerance species will be in seedling recruitment (Webster & Lorimer, 2005). Less shade-tolerant species generally require larger gap sizes for seedling regeneration, and on a larger-scale increase or keep species diversity (Lhotka et al., 2018). Some evidence indicates that even in larger individuals (>25cm diameter at breast height) the more shade tolerant species respond greater to release from local competition and the creation of canopy gaps (Jones et al., 2009). Shade tolerance of a species tends to remain the same throughout the individual's lifetime, meaning that species that are highly shade tolerant as seedlings are likely to remain shade tolerant as adults (Valladares & Niinemets, 2008).

However, gap dynamics are not uniform in time or even across stands. Gaps, even those created by natural disturbances are shown to decrease in size over time, and the previously mentioned characteristics revert back to the previously closed canopy conditions (Muscolo et al., 2014). Additionally, these conditions and the future recruitment of trees of different shade tolerances are dependent not only on the size of the gap, but the height of the canopy surrounding the gap. The effective size of these gaps is decreased as the surrounding canopy becomes taller (Lhotka et al., 2018).

Gap sizes, and the creation of them is often the starting point for land managers when looking to enhance ecosystem services and for timber production and sale. This generalized approach tends to result in the more 'intense' forest management strategies (e.g. clear-cutting), as a result of this two pronged approach, which results in poor outcomes from this increase in gaps, and increased gap size (Kern et al., 2017). Negative outcomes from this include, absent of any tree regeneration, lowered species diversity in these new gaps, and stratification along light and moisture gradients as a result of the gap creation (Kern et al., 2017). Group selection cuts can have regeneration failures that occur when advance regeneration (sapling population in the understory) is lacking and requires more land management (e.g. herbicide application, scarification) (Knapp et al., 2021). Individual selection cuts, aimed to reduce the effects of group selection, results in skid trails (temporary trails for equipment travel) and has been shown to have no increase in growth of residual trees (Moreau et al., 2020).

As forested areas and plantations used for timber harvesting are in high demand for not only timber production but other ecosystem services (e.g. biodiversity, carbon storage) increases land managers have to account for a wider variety of management outcomes (Gustafsson et al., 2012). This has caused a shift away from more impactful forest management strategies (e.g. clear-cutting) and instead a shift towards less intense strategies (e.g. group or single tree selection). Included in less intense strategies, other strategies to emulate more natural, undisturbed forest conditions are proposed, as a way to increase species diversity and enhance ecosystem resilience (Hupperts et al., 2019). The perhaps most common method includes the creation of coarse wood debris, that results in increased structural heterogeneity through standing dead wood (Harmon et al., 2004). The creation of standing dead wood, including snags and dead fallen trees is

In contrast to other management options, girdling is a process that is intended to promote ecosystem functions and does not give the option for harvesting of existing boles (Noel, 1970). Typically characterized as a more moderate or less intense treatment that emulates natural disturbances as compared to more severe, human-caused disturbances like clear cutting (Hupperts et al., 2019). A marked difference between girdling and other management practices is that girdling causes a slow death of the tree. While a tree may not fully die after one season, by the second season the tree will have fewer leaves than a healthy individual (Noel, 1970). For trees to fully die it can take an average of 4.5 years since girdling treatment (Fassnacht & Steele, 2016). As such the full effects of girdling treatment could take several years before the effects are seen.

Girdling, and other management effects that result in the creation of woody debris, have been shown to result in other ecosystem benefits. The addition of dead wood in ecosystems has been found to increase species diversity particularly of saproxylic insects and fungi (Sandström et al., 2019). Additionally, the presence of dead woody debris allows for important habitats for a variety of forest organisms (e.g. birds, mammals, fungi). (Keddy & Drummond, 1996). Additionally, the presence of woody debris helps to increase recruitment of more shade intolerant species (e.g. swamp birch) in canopy gaps (Bolton & D'Amato, 2011).

However, as is similar with the role of canopy gaps and shade intolerant recruitment, much of the effects on girdling are unclear and studies vary greatly in duration (Bolton & D'Amato, 2011; Fassnacht & Steele, 2016; Gough et al., 2020). As land manager goals have broadened, to enhance ecosystem services and forest health, while in certain cases providing timber value, girdling as a forest management treatment has become more appealing.

CHAPTER II

FOREST CANOPY STRUTURE'S RESPONSE TO LESS INTENSE MANAGEMENT

Introduction

Canopy structure, as indicated by metrics like canopy openness and leaf area index (LAI), is an important feature of forests that influence their functions (Woodgate et al., 2015). For example, the structure of the canopy has direct influences on the amount of light that can penetrate the lower levels of the forest and as a result of this, will have influence on the forest floor community (Hardiman et al., 2013; Jennings et al., 1999). Canopy structure can also influence soil chemical, physical and biotic properties (Mueller et al., 2016; Muscolo et al., 2014). Consequently, the shifts in canopy structure induced by forest succession, disturbance, and management are important to accurately monitor and understand. For example, losses of forested land coupled with recent reforestation, and changing land management practices, has created a greater need to understand how canopy structure in early and middle successional forests is evolving and can be optimally managed to enhance ecosystem services (Fahey et al., 2016).

A legacy of different forest management practices has partly illustrated the various ways in which different cutting and harvesting methods influence canopy cover. For example, a more 'open' canopy with enhanced understory light availability in large forest patches has often been achieved using gap-based approaches in which large gaps (>0.2 ha) of similarly aged individuals are harvested (Webster & Lorimer, 2005). In some places/contexts (eg. Northern hardwood forests of Quebec, Canada (Forget et al., 2007)), 'selection cutting' (of individual trees) has been encouraged as a more moderate and heterogenous alternative to the commonly used gap based harvesting methods (Kern et al., 2017). When compared to plots that receive no treatment, plots with individual selection cuts show a greater canopy openness. This increased canopy openness declines with time since cutting (Beaudet & Messier, 2002).

Tree girdling can also be used as a moderate disturbance that closely resembles the influence of pest insects (Grigri et al., 2020). As a forest management practice, girdling is slower, less rapid change in canopy cover compared to historically common harvesting methods, like gap-based approaches or selection cutting in a study of 756 individuals, 75-85% had died between 4 and 5 years of treatment (Fassnacht & Steele, 2016).

Girdling has been used in efforts to advance young or even-aged forests into late successional, uneven-aged stands. This more moderate disturbance is used in land management practices that seek a more ecological management approach in attempts to increase heterogeneity in the composition and structure of forest stands (Fahey et al., 2016). The amount of girdling that is applied at the stand level has shown to have varying influences on canopy cover. For example studies in which early successional species were girdled to promote mid to late successional growth resulting in 39% of the total basal area experiencing girdling saw a 44% decrease in LAI two years post treatment (Gough et al., 2013). Stand level decrease in LAI has been shown to require severe disturbance severity levels, as much as 85% (Grigri et al., 2020). In both of these studies the large scale girdling was applied in attempts to mimic natural disturbances like insect outbreak. Due to this experimental design, girdling was performed in a way that does not give targeted release to individual trees. The effects of girdling specifically for forest management purposes remains largely untested (Gough et al., 2020).

Assessment of forest canopy structure is a complex and ever changing challenge and currently there exist several different ways to assess the canopy cover and structure, each of which provides a separate metric that has associated advantages and disadvantages. Despite the variety of measurement options, the forest canopy and different metrics associated with measuring canopy remain important tools to assessing the forest conditions (Woodgate et al., 2015). Various measures including litter traps, hemispherical photography, and indirect light measures can be used together or solely to illustrate the dynamics of canopy cover (Stuart-Haëntjens et al., 2015; Woodgate et al., 2015). Leaf Area Index (LAI) is one such metric that is a common and widely used metric which provides a measure of the leaf surface area of the surrounding canopy (Woodgate et al., 2015). Canopy openness (CO) is another measure used similarly for assessment of understory light conditions (Beaudet & Messier, 2002).

Litter traps are used as a more direct measure of LAI for forests dominated by deciduous trees ((Burton et al., 1991). The traps can be harvested for assessment of total leaf area and can even be separated by species as well (Gough et al., 2013). Despite litterfall collections having a higher accuracy than allometric assessments, this metric can only be assessed at the end of the growing season and therefore does not account for temporal assessments (Cutini et al., 1998). Additionally, the assessment of litter fall traps is very labor intensive and to ensure accurate measurements requires many replicants (Chason et al., 1991). Estimates of LAI derived from litter traps are also less informative of canopy structure than other measures, because other important elements of the tree canopy are not measured (e.g., branches) (Ariza-Carricondo et al., 2019; Dufrêne & Bréda, 1995; Rhoads et al., 2004).

Indirect methods are also used as a more common and less labor intensive ways of measuring the canopy structure and cover. A method used in increasing frequency is digital hemispherical photos of the forest canopy (Macfarlane et al., 2014). Hemispherical photography typically requires use of a commercially available digital camera and fish eye lens aimed at the forest canopy (Beckschäfer et al., 2013). This method is useful for providing both LAI and canopy openness (Pearse et al., 2016). However, despite hemispherical photography's usefulness is assessing forested light conditions there exist a variety of ways to process the photos, and procedures are rarely standardized (x`. Moreover, the process of collecting canopy cover and LAI data from hemispherical photos has many areas in which subjectivity can cause disparities between measurements (Beckschäfer et al., 2013). Of large importance is the binarization of these digital images, commonly done by manually setting a threshold for classification of pixels into sky or canopy classifications, has been adapted to automated algorithmic applications. Given the variety of software available to collect data, there exist a variety of algorithms present, however there are several algorithms that exist to accurately binarize hemispherical photos (Glatthorn & Beckschäfer, 2014). Due to this time consuming and subjective manual process the development of automatic algorithms has become much more prevalent. Compared to other indirect measures of canopy structure (see below), the two-dimensional nature of hemispherical imagery results in courser or less accurate descriptions of three-dimensional tree canopies (Rich, 1990).

Optical light instruments have been used for several decades now as a way to obtain LAI measures for a variety of plant community types (Pearse et al., 2016). The most common instrument used is the LI-COR LAI-2200 (LI-COR, 2012). The main mechanism behind this method of data collection involves the difference of light transmittance between above canopy readings and below canopy readings (Pearse et al., 2016). With the increasing commonality in usage there exist some large concerns in the usage of such instruments such as the dependency upon specific sky conditions (Pearse et al., 2016). Furthermore measurements made in this method may not distinguish woody material from leaf tissue, and when compared to more direct methods, LAI, tends to be overestimated (Ariza-Carricondo et al., 2019). Despite these challenges in the operation of this instrument, it can used at the stand or transect level with ease and in a nondestructive manner. As with hemispherical photography, many of the LAI measures made with instruments like the LAI 2200 are not often applied at the individual tree level to access how conditions around the target tree change.

This study is designed to address how girdling influences canopy structure in a secondary growth, temperate forest. Additionally, it aims to use LAI measurements to assess these canopy changes in a more targeted approach to evaluate the neighborhoods around individual trees. Understanding how targeted girdling treatments influences individual trees and how the light conditions change around each tree will provide useful information to land managers seeking to create better conditions for secondary forests that are prone to extreme management challenges. We expect that targeted girdling will result in decreased canopy cover and LAI with increasing changes in time since release. We also expect to observe greater increases in canopy openness and understory light availability, and larger decreases through LAI measurements, in tree neighborhoods that experienced more intense levels of girdling. Lastly, we expect that metrics of canopy structure and light availability derived from the LI-2200 will detect larger treatment effects, and earlier, because this method has the potential to detect changes to canopy structure in three dimensions (compared to two dimensions observed by hemispherical photography). Given tremendous heterogeneity in tree stands and neighborhoods within forests, we will also explore the extent to which girdling effects are dependent on characteristics of our sampling neighborhoods, including the size and species of the focal tree within the center of each neighborhood.

Methods

Site Description and Experimental Design

This site was located within the area of the Working Woods at Holden Arboretum in northeast Ohio. This site was previously agricultural land that has been abandoned for around 60 years. The area is young forest that is dominated by red maple and tulip poplar in the overstory and red maple and sugar maple in the understory. This area was divided into nine plots (100 m x 100 m each) that were further divided in half to apply treatments in separate years. The plots were evenly divided among three treatments, including control plots with no management, plots where improvement cuts (IC) were made to remove poor quality or undesired trees, and plots where improvement cuts and timber stand improvement (IC+TSI), or the removal of invasive shrubs and grape vines occurred. Plots were designated by their location in relation to the larger area, the specific treatment and the year it was applied and are shown in Figure 2.1. A total of 457 trees were identified and tagged to measure a variety of metrics within the focal plots. Trees were selected that were midstory or understory trees that experienced a large amount of competition from neighboring trees. Trees were also selected that were mostly red maple, sugar maple and tulip poplar. Small understory elms, sugar maples and red maples were selected as well. The initial size of these trees ranged from 1.6 to 62.8 cm (diameter at breast height). Most target trees were red maples (n=137), sugar maples (n=170), tulip poplars (n=95), red oak (n=12), or elms (n=40).

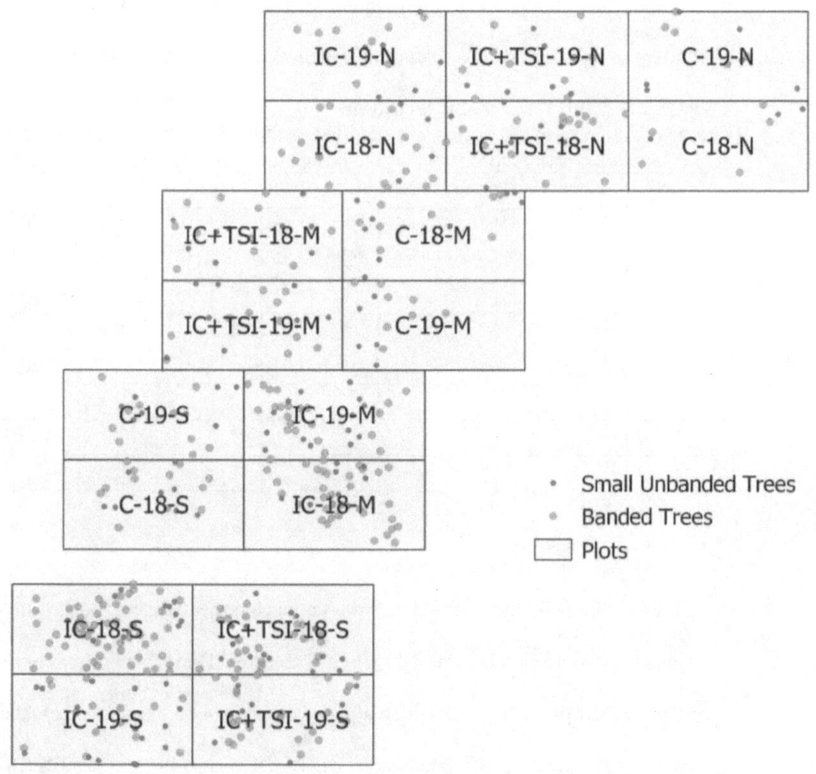

FIGURE 2.1 Map of the study site featuring target trees.

Plot layout with all target trees both banded and unbanded trees are shown. Plots are designated by treatment, year of treatment and the direction of the plot's layout.

Treatments were applied either during the winter of 2018 or the winter of 2019. Trees were girdled that were in poor health or form (e.g. multiple stems), were undesirable species, or individuals that were too close to target trees. Most of the trees that were selected for removal were girdled, but those that were too close to target trees or public trails were felled and left on the forest floor in place. Table 2.1 describes the population of trees that received 'cutting' (girdling or felled) treatment.

TABLE 2.1 'Cut' trees characterization.
Characterization of the trees that were girdled or felled in both years of the treatment application. Values are represented as the basal area of the tree that is cut (cm^2).

	2018 Cut Basal Area			2019 Cut Basal Area		
	Girdled	Felled	Total ('Cut' Value)	Girdled	Felled	Total ('Cut' Value)
Mean	1272.7	936.4	1236.4	1113.2	967.4	1098.3
Standard Deviation	1221.8	919.9	1195.4	769.6	648.5	757.7
Maximum	9746.8	2700.5	9746.8	4738.7	2811.3	4738.7
Minimum	98.5	89.9	89.9	116.9	167.4	116.9
N	165	20	185	166	19	185

GIS

The latitude, longitude, and elevation of each target tree and each girdled or felled tree was recorded using a Garmin GLO 2 GPS receiver. These locations were recorded during the winter of 2019-2020, which allowed for improved signal strength and accuracy due the senescence of deciduous leaves. QGIS was used to identify 10 m radii around each target tree (QGIS Development Team, 2021). The number, species, and size of each girdled or felled tree within each 10 meter radius was used for determining release from competition.

LAI 2200 Measurements

Leaf area index (LAI) values were collected using a LI-COR 2200C Canopy Analyzer for 356 target trees in 2019 and 388 target trees in 2020 and 387 target trees in 2021 (LI-COR, 2012). LAI was not measured effectively or comprehensively in 2018 due to time and labor limitations. LAI measures in 2019 were made between mid-September and early October. LAI measures in 2020 were made between mid-July and late August. LAI measures in 2021 were made between early July and mid-July. The 2020 and 2021 measurements were only taken during mostly clear sky conditions. LAI measurements were taken in each cardinal direction around a target tree similar to López-Serrano et al., method (2000). Each wand had the 90° view cap attached to exclude the trunk of the target tree and the operator from the view of the sensor. Due to the generally closed nature of the canopy, one wand for above-canopy (open sky) measurements was set in a herbaceous field 20 meters outside the edge of the Working Woods citizen science trail head. The second wand was attached to a monopod and used to take below canopy readings at each cardinal direction at the target tree. Measurements were made 2 m away from each target tree looking in the cardinal direction. Low shrub or other foliage cover was displaced to allow for an unobstructed view of the canopy. In 2019, measurements were repeated three times at each cardinal direction. Repeat measures of the cardinal directions were found to have little variation and as a result the 2020 and 2021 measures were only done once in each cardinal direction. Readings were compiled and LAI and DIFN values calculated through the FV2200 software (LI-COR, 2012). DIFN values are representation of the canopy structure and diffuse short wave length light that is getting through the canopy. This value provides a more direct value of what foliage is blocking the understory light and does not account for the orientation of the leaves (LI-COR, 2012). Scattering correction files were recorded for each set of measures on days when sky conditions allowed them to collected (non-overcast). Scattering correction was not applied to LAI values as it resulted in lower correlation between repeated measures from both 2019 and 2020.

Hemispherical Photography

Hemispherical photos were taken in 2018, 2019, 2020 and 2021 at 122, 447, 349 and 378 target trees respectively within the study location. As for LAI reading, variability in the number of trees measured each year reflects variability in the time and labor available for measurements each year. Photos were taken in 2018 between late July and late August. Photos taken in 2019, 2020, and 2021 were taken between late June and late August. Following standard hemispherical photography methods, a NIKON D5600 camera was attached to a tripod to sit above the ground cover foliage (Rich, 1990). Photos were taken from below the canopy at each cardinal direction 2 meters away from each target tree.

Hemispherical photos were processed for finding canopy openness (CO) using two separate methods. Photos were processed manually using Gap Light Analyzer (GLA) (Frazer et al., 1999) where a pixel threshold was estimated by the observer. The values of this threshold determines which pixels are designated as 'open' and which pixels are designated as 'canopy'. This method will be referred to as manually processed photos. An automated process was also used due to concerns that the manual method would not result in standardized CO due to subjectivity of observer decisions about which threshold values to use, which could introduce additional error when images were processed by different observers. Prior to automated image processing, images were color-corrected using GIMP software, and the color curve of the image was adjusted to avoid overly bright pixels. The same color curve correction was applied to every photo. Color-corrected images were binarized through ImageJ with the Hemispherical 2.0 plugin which applies the minimum algorithm (Beckschäfer, 2015). This binarization process effectively eliminates the need to determine a threshold value for distinguishing between open sky pixels and canopy pixels, by finding a minimum bin of the image's gray value histogram and using that as minimum as the threshold value (Glatthorn & Beckschäfer, 2014). Binarized images were then processed through GLA similarly to the manual method, but without application of a threshold value. These values were averaged over the four photos to provide CO for each target tree.

Due to the concerns over standardized threshold applications canopy openness values from both methods were tested against their methodological counter parts and their future measurements (Table .12). Automated photos, when compared with their manual counter parts were closely related (average R2=0.68, $p<0.001$ for all 3 years). Manually calculated values tended to have greater variability, with less significance between similarly calculated values from different years, than the automatedly calculated values (manual avg R^2=0.11, automated avg $R^2 = 0.14$). For these reasons, and concerns over the subjective nature of manually processed photos, the canopy openness value from the automated process was used in this study. Initial values of automated canopy openness are shown in Table 2.3.

TABLE 2.2 Univariate analysis between canopy openness generated from automated and manually processed hemispherical photos.
Significant p values are denoted with asterisks. *** < 0.001, **<0.01. *<0.05.

	Auto 18	Auto 19	Auto 20	Manual 18	Manual 19	Manual 20
Manual 18	0.82 ***	0.21 ***	0.10 ***	NA	0.25 ***	0.073 *
Manual 19	0.25 ***	0.81 ***	0.044 ***	0.25 ***	NA	0.0043
Manual 20	0.10 **	0.0037	0.42 ***	0.073 *	0.00043	NA
Auto 18	NA	0.28 ***	0.077 **			
Auto 19	0.28 ***	NA	0.069 ***			
Auto 20	0.077 **	0.069 ***	NA			

TABLE 2.3 Initial conditions from the first measure of each canopy metric. Measurements are separated into the plot designation, and by the categorization of whether the trees experienced cutting in their 10 m radius and when the cuts took place in reference to 2020 measures (cut category).

Treatment	**Canopy Openness (%) in 2018**				
	Mean	**Standard Deviation**	**Maximum**	**Minimum**	**n**
C	**17.32**	**7.00**	**32.38**	**8.17**	**38**
IC	**20.62**	**5.6**	**36.30**	**10.89**	**37**
IC+TSI	**20.12**	**5.96**	**33.15**	**11.21**	**38**
Total	**19.34**	**6.34**	**36.30**	**8.17**	**113**
Cut Category					
C	**17.74**	**6.90**	**32.82**	**8.17**	**43**
IC-0.5	**22.10**	**4.82**	**33.15**	**11.47**	**34**
IC-1.5	**18.36**	**6.27**	**36.30**	**10.89**	**32**
IC-1.5+0.5	**21.05**	**5.79**	**26.64**	**12.44**	**4**

Neighborhood Competition Indices

In order to assess the level of competition from other trees around each target tree, several neighborhood competition indices (NCI) were calculated. The calculations were based on Hartmann and Messier's (2011) models which factored in size of the competitor, and distance of target tree to the competitor. The general model for such neighborhood competition indices is shown below. NCI models were used to assess both the amount of release target trees were experiencing as a result of girding and felling ('cutting'), as well as the amount of living competition each tree experienced. Both cut NCIs and living NCIs were assessed for a 10 meter radius around each target tree (Looney et al., 2018).

$$NCI_i = \sum_{j=1}^{N} \frac{(DBH_j)^\alpha}{(dist_{ij})^\beta}$$

This model is an example of general model where DBH_j is the dbh of a girdled or living tree *j* that is a distance represented by $dist_{ij}$ from the target tree, represented by *i*. The value of α was either 0, 1, or 2 depending on the model, meaning competitors were counted (α=0), the dbh of competitors summed (α=1), or the basal area of competitors summed (α=2). The value of β was either 0, 0.5, 1, or 2, meaning the distance was not included (β=0), the square root of distance of each competitor was used (β=0.5), the non-transformed distance (β=1), or the distance squared (β=2). Thus for a β of 1, the effect of a competitior decreases as a simple linear function of its distance from the target tree, wile for β of 0.5 the effect of a competitor decreases slower with distance (compared to 1), and or a β of 2 the effect of a competitor decreases faster with distance (compared to 1).

Health Assessment

In 2021, a health assessment was performed for a subset of the target trees, focusing on the larger trees (n=286). This health assessment featured a canopy assessment as well as tree health/damage and site characteristics around each target tree. Canopy measures included canopy position (canopy position in relation to the midpoint of the surrounding overstory), canopy light exposure (the amount of sides of the canopy that have more than 30% of that side exposed to unobstructed light) and crown class (classification based on canopy position and light exposure). These were measured following standard forestry practices (Bechtold, 2003). Additionally included in the canopy assessment was the number of live and girdled overtopping trees at each target tree canopy. Trees were counted as 'overtopping' if they had branches that reached over any part of the canopy of the target tree. Health-related metrics that were measured included crown dieback (percentage of the canopy that has dead branches), root flare (presence/absence of taper on the trunk of tree), lower trunk damage (presence/absence of any peeling, broken bark), other damage to the trunk (presence/absence of bark damage, fungus or insect wounds, etc.), bare soil ground cover (percentage of ground under canopy without leaf litter/other decomposed organic material), microtopography assessment (categorical assessment of adjacent areas that contributes to or receives run off), and overall tree condition (categorical assessment of health of tree by canopy vigor, growth form and absence of dead branches). These indices of tree health were measured following the protocols established by Vogt et al (2014).

Statistical Analysis

Overview. The statistical analyses were executed in three phases. In the first phase of analyses, the separate (univariate) effects of a variety of ecological and environmental variables on metrics of canopy structure were assessed using simple regression or ANOVA models. Pairwise relationships among the different yearly metrics of canopy structure (e.g., canopy openness, LAI, and DIFN) were also assessed. In the second phase of analyses, two-factor ANOVA models were used to assess the effects of treatments in models that included the previous year's measurements. In the third phase of analyses, linear-mixed models were used to simultaneously assess the effects of multiple variables on metrics of canopy structure. Dependent and independent variables were transformed using a log-function or square-root function if those transformations resulted in a more normal distribution.

Univariate analysis. Simple (univariate) linear models were used to assess the effects of a variety of ecological variables on the various metrics of canopy structure, to describe the correlations among the different metrics of canopy structure, and to describe the correlations between earlier and later measures of canopy structure (measurements made from one year to another) potential spatial covariates (latitude, longitude, elevation), various metrics of tree and a tree's position in the canopy, and alternative NCIs representing the number, size, and proximity to cut or living trees in a 10 meter radii around each target tree.

Several different variables were used to describe whether, and to what extent, each targe tree 'experienced' the IC treatment. First, each target tree was separated into a category, which designated whether a target tree had experienced cutting within their 10 meter radius, and how many years had passed, since the cutting was implemented. Category designation included "C" for trees that did not experience any cutting, "IC-0.5" for trees that had cutting the winter prior that year's measures of canopy structure, "IC-1.5" for trees that had cutting two winters prior to that year's measures of canopy structure and "IC-0.5+1.5" for trees that had cutting occur both one and two winters prior to that year's measure of canopy structure. The effects of NCI from trees that 'experienced' a release from competition (i.e. any trees without cutting in a 10 meter radius were not included).

The effect of each independent variable was tested in separate linear models of each metric of canopy structure. The *lm* function in the R software (version 1.2.1335) was used to create these statistical models and corresponding model statistics (R Core Team, 2019).

Two-factor models. For each canopy structure metric, treatment effects were assessed in a model that included the previous year's canopy structure measures, a variable describing either the cutting treatment or intensity of the cutting, and an interaction term between cutting treatment/intensity and the previous metric measurement. These models allow for effects of treatments to be interpreted as a treatment-induced shift in canopy structure, relative to canopy structure measured the previous year. This model was used due to the large variation of canopy structure within the plots and the potential for pre-treatment canopy structure to be heterogenous along treatments and plots and variable intensities of intensity. Two-factor models were specified with the *lm* function, and summarized with the *Anova* function from the car package in the R software (Fox & Weisberg, 2019; R Core Team, 2019).

Mixed effects linear models. A subset of potential predictors including variables from the univariate analysis were assessed in linear mixed-models. Predictors were included in these models if: *i)* the results of the simple ANOVA or regression models showed evidence for influencing canopy structure, *ii)* the predictors were useful for assessing how these metrics might be similar along a spatial gradient (e.g. latitude, longitude, elevation of each target tree), or *iii)* if they were important to testing original hypotheses (e.g. species). Like the two-factor models described above, each mixed-model including a measure of the same canopy structure from the previous year to ensure treatment effects were accounting for pre-existing variability. This allows for effects of treatments to be interpreted as a treatment-induced shift in canopy structure, relative to canopy structure measured the previous year. A taxonomic variable, indicating the species of the target tree was included as a random effect to account for potential differences in species allocation of biomass, leaf sizes, leaf angles and area which could influence these canopy structure metrics. Mixed models were specified using the lmer function from the lme4 R package (Bates et al., 2015). The effects of each model predictor were assessed using Type II ANOVA.

Temporal patterns in canopy structure. To assess temporal patterns in canopy structure, a subset of 204 target trees that had three years of canopy metric measurements that ranged from pretreatment (-0.5 years prior to treatment), the year immediately following treatment (0.5 years post treatment), and two growing seasons post treatment (1.5 years post treatment). In addition to the time relative to cuts, the trees were organized by whether they received treatment or not, either "C" for control trees with no treatment and "IC" for trees that had cutting within a 10 meter radius. Canopy openness from automated hemispherical photos (n=52, for C, n=152 for IC), and DIFN and LAI from LI-COR 2200C (n=39 for C, n=104 for IC) were all assessed across the above temporal range. For canopy openness, trees in 2018 plots had their 2018, 2019 and 2020 values used and trees in 2019 plots had their 2019, 2020 and 2021 values used. For LAI and DIFN, only trees in 2019 plots were used and the values from 2019, 2020 and 2021 were used. Means and standard deviation were calculated for each year and compared qualitatively to examine if there were any differences in the canopy structure metrics over time.

Results

Univariate Analysis

Pre-treatment variability. It is clear that latitude and longitude have an effect on the initial canopy conditions, as well as the amount of live trees overtopping the target tree. Specifically, canopy openness and DIFN decreased with latitude (CO in 2018: $R^2 = -0.35$, $p<0.001$. DIFN in 2019: $R^2 = -0.18$, $p<0.001$), while LAI increased with latitude ($R^2 = 0.20$, $p<0.001$, Table 2.4). The future cut categorization showed that there was pre-existing variability between these categories for the canopy openness in 2018 ($R^2 = 0.092$, $p<0.05$). In 2018, before any tree was girdled or felled, canopy openness was lower, for trees that would become categorized as C (mean CO = 17.7%) and IC-1.5 (mean CO = 18.4 %) compared to trees categorized as IC-0.5 (mean CO = 22.1%) and IC-15+0.5 (mean CO = 21.0 %, Table 2.3). However, cut intensity for the trees later experiencing cutting did not have an effect on the initial canopy structure ($R^2 = 0.076$, $p > 0.05$). There was also no apparent correlation with canopy structure for other environmental or spatial variables, including elevation, dead/girdle overtopping, crown metrics, and living basal area around the target tree (Table 2.4).

Covariance among different metrics of canopy structure and measurements over time. Each dependent variable is closely related to the other dependent variables as well as the same variable measured at different years (Table 2.4). The weakest relation is between the canopy openness in 2020 and 2019 ($R^2 = 0.07$, $p<0.01$). The least statistically significant correlation was between the canopy openness in 2020 and 2018 ($p = 0.0034$). The strongest relation was between LAI in 2019 and LAI in 2020 ($R^2 = 0.35$, $p<0.001$). However, previous year measures are a significant predictor for all dependent variables of the following year (all p values < 0.001). LAI measures were negatively correlated with the DIFN measures made in the same year, for both of the years LAI and DIFN were measured, which is to be expected ($R^2 < -0.87$, $p<0.001$, Table 2.4).

Covariance of canopy structure in 2019 and 2020 with environmental, ecological, and treatment-related variables. Nearly all metrics of canopy structure show relation to the latitude at each target tree (p value <0.05 for all canopy metrics). The longitude was a significant predictor for canopy openness in 2018 and 2019, and LAI and DIFN in 2019 (p values < 0.001). Elevation was also significantly related to both LAI and DIFN in 2020 (p values 0.001 < 0.01). Live overtopping also was related to pretreatment canopy cover as well as LAI in 2019 and DIFN in both 2019 and 2020 (p value <0.05 for LAI 2019, DIFN 2019 and 2020. p value <0.001 canopy openness 2018). None of these predictors explained more than 40% (range of 40% - 1.4 %) of the variability in canopy structure metrics (Table 2.4), despite the significant p values.

The LAI and DIFN measures in 2020 were significantly related to the cut category, with the LAI in 2020 being only slight more related ($R^2 = 0.084$, $p<0.001$) than the DIFN in 2020 is related ($R^2 = 0.057$, $p< 0.001$). Cut category determined by 2019 conditions is also a good predictor for LAI and DIFN in 2019, where DIFN was positively related ($R^2 = 0.027$, $p<0.01$) and LAI negatively related ($R^2 = -0.020$, $p<0.01$). Cut category in 2020 is more closely related than these canopy structure metrics but is not as significant as them ($R^2 = 0.092$, $p<0.05$).

The different distance dependent cut area NCIs were found to be strong predictors for 2020 LAI and slightly less strong predictors for DIFN in 2020 (p value <0.001 for LAI, <0.05 for DIFN). Trees with higher NCI values, i.e., which experienced more girdling and/or felling within a 10 m radius, tended to have higher values of LAI (lower for β=0 and 2 NCI), and higher values of DIFN (Table 2.4). However none of the distance dependent were more strongly related with any dependent variable and NCI metrics explained relatively little variation in canopy structure (R^2 from 0.076 to 3.867 x 10^{-5}). Living NCI was a good predictor for only the LAI and DIFN measures in both 2020 and 2019 (all p values < 0.05).

TABLE 2.4 Covariate and variable assessment on dependent canopy structure variables.
Independent variables that are tested against themselves are left blank." R squared values are reported with direction relation shown as well. Significant p values are denoted with asterisks. *** < 0.001, **<0.01. *<0.05.

Variables	% Canopy Openness 2018	% Canopy Openness 2019	% Canopy Openness 2020	LAI 2019	LAI 2020	DIFN 2019	DIFN 2020
% Canopy Cover 2018	-------	0.28***	0.077 **	-0.48 ***	-0.27 ***	0.38 ***	0.27 ***
% Canopy Cover 2019	0.28 ***	-------	0.069 ***	-0.25 ***	-0.23***	0.22 ***	0.17 ***
% Canopy Cover 2020	0.077 **	0.069***	-------	-0.13 ***	-0.24 ***	0.12 ***	0.21 ***
LAI 2019	-0.48 ***	-0.25***	-0.13***	---------------	0.35 ***	-0.93 ***	-0.29 ***
LAI 2020	-0.27 ***	-0.23***	-0.24***	0.35 ***	---------------	-0.33 ***	-0.87 ***
DIFN 2019	0.38 ***	0.22***	0.12***	-0.93 ***	-0.33 ***	---------------	0.29 ***
DIFN 2020	0.27 ***	0.17***	0.21***	-0.29 ***	-0.87 ***	0.29 ***	---------------
Latitude	-0.35 ***	-0.020**	-0.012*	0.20 ***	0.057 ***	-0.18 ***	-0.033 ***
Longitude	-0.41 ***	-0.054***	-0.0076	0.22 ***	0.0028	-0.18 ***	-0.00011
Elevation	0.0084	0.020**	0.00029	-0.0082	0.019 **	0.0018	-0.022 **
Initial Size dbh18	-0.0015	-0.0047	-0.0079	-0.00042	0.0014	0.0033	-0.0002
Species	0.0050	0.041**	-0.016	0.017	0.013	0.014	-0.0047
Crown Position	-0.00031	0.0007	0.0088*	0.0013	-0.0018	-0.00038	-5.08^-5
Crown Light Exposure	0.028	1.416^-5	-0.0015	-0.017*	-0.0050	0.016	0.014*
Canopy Class	-0.023	0.00095	0.0044	0.0039	-0.00058	-0.0014	-0.0020
Live Overtopping	-0.079**	-0.00083	0.00054	0.016*	0.0046	-0.012*	-0.0085
Dead/girdle Overtopping	0.0015	0.023**	6.21^-5	0.00036	-0.0052	-0.0014	-0.00015
Crown Dieback	-0.0025	0.0049	-0.0022	-3.37^-5	4.551^-6	4.91^-9	-0.00028
Living Basal Area	0.00010	-0.0047	-0.0068	0.046 *	0.032 *	-0.064 **	-0.028 *
Cut Category	0.092 *	0.0047	0.013	-0.020**	0.084 ***	0.027 **	0.057 ***
Cut Intensity β=0	0.00079	0.020 *	0.0031	0.0025	-0.025 **	3.87 x 10^{-5}	0.012
Cut Intensity β=0.5	.020	0.020 *	0.0050	0.0020	0.044 ***	0.00036	0.018 *
Cut Intensity β=1	0.048	0.015	0.0050	0.0018	0.050 ***	0.00059	0.018 *
Cut Intensity β=2	0.076	0.0048	0.0040	0.0026	-0.044 ***	0.00027	0.015 *

Two-Factor Models

The two-factor models of every dependent variable that has a previous year measurement is represented in table 2.5. Target tree sample size in the model is also reported for each model. From this analysis, cut category and intensity alone are good predictors for only LAI measured in 2020, and cut category was a good predictor for DIFN 2020. All previous measures were good predictors for each of these dependent variables ($p < 0.001$), with the model measuring cut intensity and canopy openness in 2019 showing the least strength ($p < 0.01$). The only interaction term that acts as a good predictor is in the model canopy openness in 2020 that includes the cut category.

The relation between previous measurement and the treatment category for each dependent variable is shown in Figure 2.2. The differences between categories in canopy openness is clear within graph A showing the canopy openness measured in 2020. Treatments IC-0.5 showed a significant difference ($p < 0.001$) to the trees that have not experienced any cutting (treatment C). The other treatments (IC-1.5 and IC-1.5+0.5) showed different trends but were not significantly different from nontreatment trees. None of the other dependent variables showed any significant differences between treatment categories and the previous year's canopy structure measurements.

TABLE 2.5 Two-factor models of canopy structure
showing the p-value between each dependent and cut category/intensity, previous measures and the interaction between them. Asterisks included for emphasis *** < 0.001, **<0.01. *<0.05.

Variables	Canopy Openness 2020		LAI 2020		DIFN 2020		Canopy Openness 2019	
	Cut category	Cut intensity	Cut Category	Cut Intensity	Cut category	Cut Intensity	Cut category	Cut Intensity
Treatment	0.12	0.52	<0.001 ***	0.0048**	0.0031 **	0.059	0.058	0.11
Previous Measures	<0.001***	<0.001***	<0.001 ***	<0.001 ***	<0.001 ***	<0.001 ***	<0.001 ***	0.00186 **
Treatment * previous measures	0.030 *	0.20	0.55	0.50	0.86	0.76	0.29	0.85
n	425	322	340	252	340	252	113	36

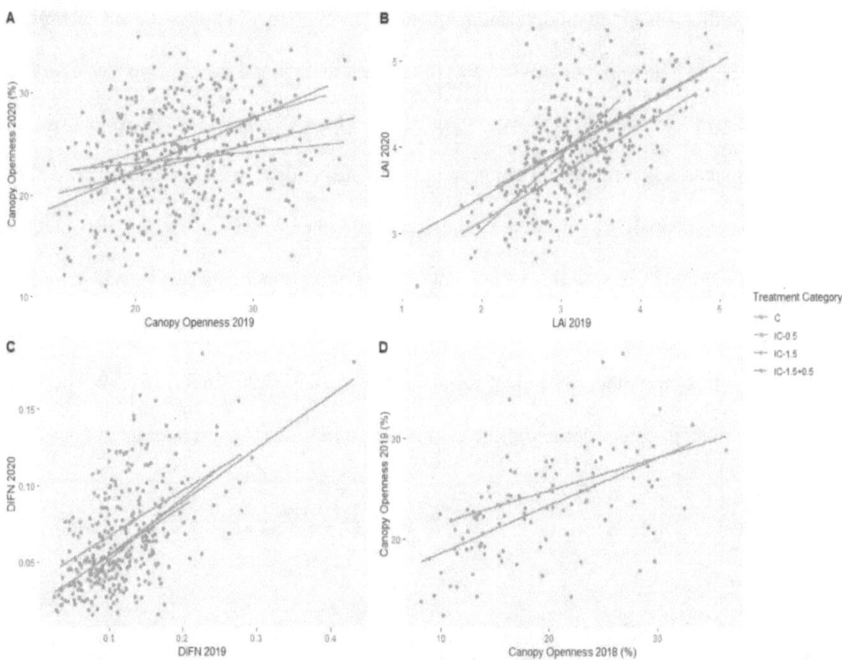

FIGURE 2.2 The two factor models of canopy structure

Models showing the relationship between the treatment category, the previous year's canopy structure measurements and A) canopy openness 2020, B) LAI 2020, C) DIFN 2020, D) canopy openness 2019.

Mixed Effects Linear Models

The mixed effect linear models representing the cut category and the cut intensity are shown in tables 2.6 and 2.7 respectively. As with the two-factor models, the cut categories acted as good predictors for LAI and DIFN measured in 2020 ($p<0.001$ for LAI 2020, and <0.01 for DIFN 2020, Table 2.6), but did not act as a good predictor for canopy openness in 2020. Additionally, like the two-factor models, the previous year's measures were significant predictors for each dependent variable ($p<0.001$). Species, as a random effect, appeared important only for LAI in 2020 ($p = 0.01306$); generally little variance was attributed to this random effect across all models. Longitude also influenced both the canopy openness ($p<0.05$), and the LAI ($p<0.001$) and DIFN ($p<0.01$, Table 2.6). Species appeared to only have an influence in the LAI measures of 2020 ($p<0.05$). Live overtopping only had an influence on canopy openness in 2019. Every interaction term that live overtopping was included was not a significant predictor for any dependent variables.

TABLE 2.6 Mixed effect linear model of canopy structure and cut category results describing p-values for each model including the cut category. *** < 0.001, **<0.01. *<0.05.

Variables	Canopy Openness 2020	LAI 2020	DIFN 2020	Canopy Openness 2019
Fixed Effects				
IC cut category	0.16	<0.001 ***	0.0012**	0.13
Longitude	0.011 *	<0.001 ***	0.0035**	0.32
Latitude	0.041	0.26	0.30	0.75
Elevation	0.31	0.45	0.21	0.023*
Live Overtopping	0.27	0.99	0.75	0.036*
autoHemi19 LAI19 DIFN19 autoHemi18	<0.001 ***	<0.001 ***	<0.001 ***	<0.001 ***
Initial Size	0.55	0.32	0.74	0.12
Interaction				
Cut Category*Initial Size	0.55	0.66	0.85	0.95
Cut Category * Live Overtopping	0.19	0.41	0.73	0.67
Cut Category * Previous Canopy	0.11	0.018 *	0.067	0.24
Random				
Species	1	0.013 *	1	0.99
n	364	292	292	102

The cut intensity did not act as a significant predictor for any of the dependent variables. Additionally, all interaction terms (all of which included cut intensity) did not act as significant predictors for any dependent variable. Longitude and latitude are significant predictors for every dependent variable except for canopy openness in 2019. Elevation acted as a significant predictor only for canopy openness in 2019 ($p < 0.01$). Like the two-factor models, the previous year's measures were significant predictors for each dependent variable ($p < 0.001$). Live overtopping was not a significant predictor for any dependent variable. Species, as a random effect, only influenced the LAI 2020 measurements. None of the interaction terms served as significant predictors for any of the dependent variables.

TABLE 2.7 Mixed effect linear model of canopy structure and cut intensity results describing p-values for each model including the cut intensity. *** < 0.001, **<0.01. *<0.05.

Variables	Canopy Openness 2020	LAI 2020	DIFN 2020	Canopy Openness 2019
Fixed Effects				
Cut Intensity	0.73	0.48	0.47	0.72
Longitude	<0.001 ***	<0.001 ***	<0.001 ***	0.16
Latitude	<0.001 ***	<0.001 ***	0.0091 **	0.44
Elevation	0.042 *	0.074	0.46	0.0046 **
Live Overtopping	0.64	0.39	0.93	0.13
autoHemi19 LAI19 DIFN19 autoHemi18	<0.001 ***	<0.001 ***	<0.001 ***	0.0053
Initial Size	0.61	0.87	0.93	0.27
Interaction				
Cut Intensity*Initial Size	0.77	0.34	0.44	0.63
Cut Intensity * Live Overtopping	0.85	0.29	0.73	0.96
Cut Intensity * Previous Canopy	0.64	0.46	0.99	0.53
Random				
Species	1	0.070*	0.93	0.99
n	279	219	219	66

Temporal Patterns in Canopy Structure

The canopy openness estimated from hemispherical photos was similar from the year immediately before treatment through the two years post treatment, for both trees that did and did not receive any girdling treatment (Figure 2.3). Mean canopy openness was greater for trees receiving treatment compared to the trees that did not receive treatment for all years including prior to and after treatment; (for example, half a year before treatment 23.3% around trees with girdling nearby (IC), and 21.0% around trees without girdling nearby (C); Table 2.8). A half a year following treatment, the IC trees had a 2.2% increase in canopy openness (from 23.3% to 23.8 %; Table 2.8). Over the same time frame, the C trees had a 9.2% increase in canopy openness (from 21.0 % to 22.9 %; Table 2.8). Additionally, for both treatment categories the variability (standard deviation and range) decreased after treatment (Table 2.8). Before and after treatment, there was a substantial overlap for the distributions of canopy openness of trees with and without girdling nearby (Figure 2.3).

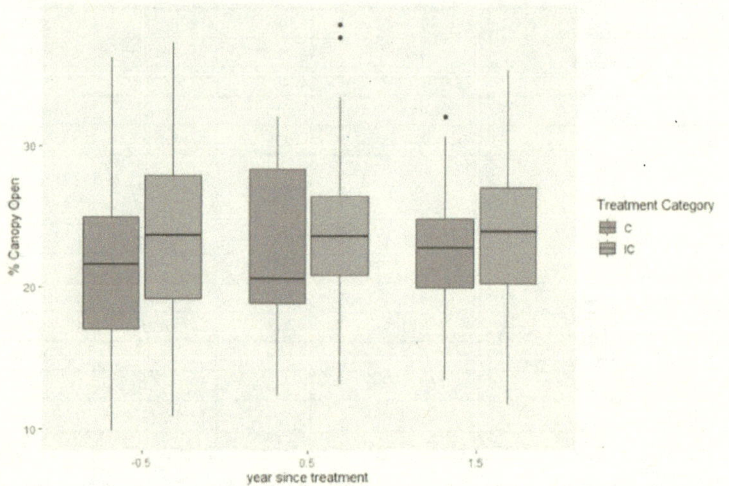

FIGURE 2.3 Temporal change of canopy openness

The LAI measurements showed an increase for both treatment categories as time passed since treatment (Figure 2.4). Trees with girdling nearby saw a 21% increase after treatment (3.1 to 3.8) and a further 4.8% increase an additional year after the treatment was applied (3.8 to 3.9, Table 2.8). The trees without girdling nearby had a 24% increase in LAI approximately 6 months after treatment (3.4 to 4.3) and a further 3% increase an additional year after the treatment (4.3 to 4.4, Table 2.8). Trees without girdling nearby exhibited a decrease in variability after treatment, while trees with girdling nearby did not see consistent changes to variability (Table 2.8). Mean values of LAI were greater for trees without girdling nearby than the trees that did have girdling nearby for all years (e.g. C=3.4, IC=3.1 for -0.5 years prior to treatment, Table 2.8). Least-squares ANOVA models showed that the differences between treatments and over time were substantial ($p<0.001$ for fixed effects for 'treatment' and 'year since treatment'), but that the difference between treatment means did not diverge substantially over time ($p=0.69$ for the interaction between treatment and year since treatment).

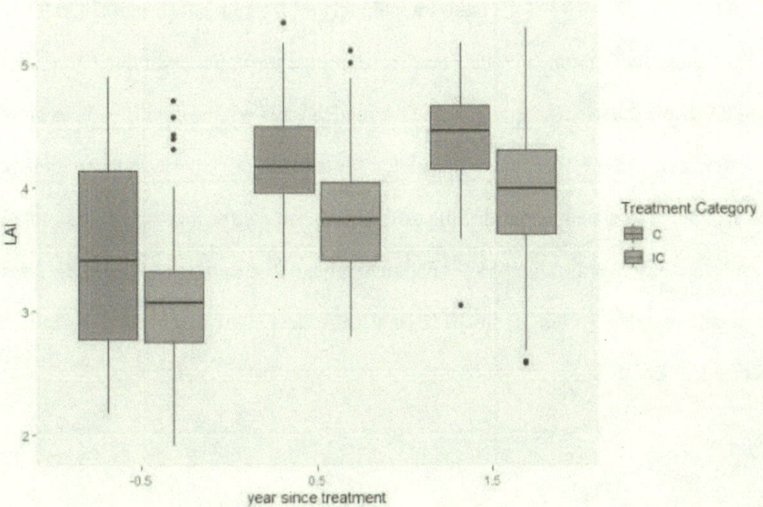

FIGURE 2.4 Temporal change of LAI

DIFN was greater around trees with girdling nearby compared to trees without girdling nearby, in all years (e.g. C = 0.10 and, IC = 0.12 prior to treatment; Table 2.8; Figure 2.5). Trees with girdling nearby showed a 37 % decrease about six months after treatment (from 0.12 to 0.07, Table 2.8) and a further 17 % decrease another year after treatment (from 0.072 to 0.06, Table 2.8). Trees that did not have girdling within the 10 meter radius saw a 47 % decrease immediately after treatment (from 0.10 to 0.05, Table 2.8) and an additional 20 % decrease an additional year after treatment (from 0.05 to 0.04, Table 2.8). Least-squares ANOVA models showed that the differences between treatments and over time were substantial ($p<0.001$ for fixed effects for 'treatment' and 'year since treatment'), but that the difference between treatment means did not diverge substantially over time ($p=0.9$ for the interaction between treatment and year since treatment). The variability in DIFN values also decreased after treatment application (Table 2.8).

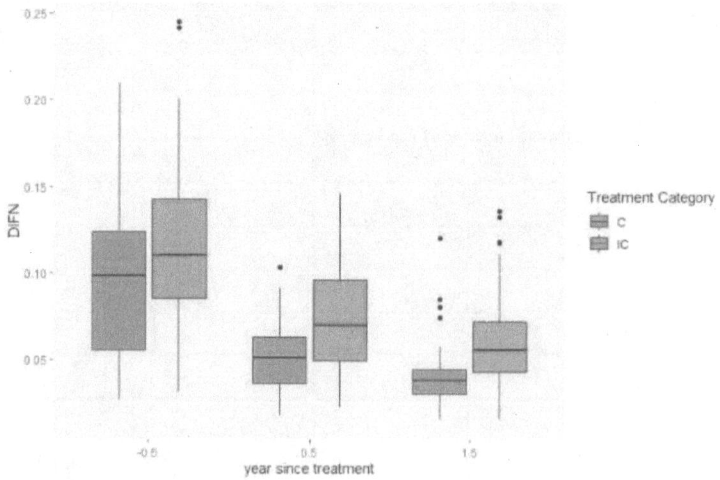

FIGURE 2.5 Temporal change of DIFN

TABLE 2.8 Descriptive statistics of the temporal change in canopy structure

			2018-2019	2019-2020	2020-2021
			Time relative to cuts (time zero)		
Treatment	Variable	Statistic	-0.5 years	+0.5 years	+1.5 years
no cuts within 10 m	canopy openness	mean	20.98	22.92	22.84
		SD	5.61	5.48	4.47
		min	9.90	12.39	13.47
		max	36.27	32.04	32.04
		range	26.37	19.66	18.57
		n	52	52	52
cuts within 10 m	canopy openness	mean	23.26	23.78	23.72
		SD	6.07	4.58	4.48
		min	10.89	13.11	11.74
		max	37.25	38.52	35.28
		range	26.36	25.41	23.54
		n	152	152	152
no cuts within 10 m	DIFN	mean	0.096	0.051	0.041
		SD	0.045	0.020	0.020
		min	0.027	0.018	0.015
		max	0.21	0.10	0.12
		range	0.18	0.085	0.11
		n	39	39	39
cuts within 10 m	DIFN	mean	0.12	0.072	0.060
		SD	0.042	0.023	0.025
		min	0.031	0.022	0.015
		max	0.24	0.15	0.14
		range	0.21	0.12	0.12
		n	104	104	104
no cuts within 10 m	LAI	mean	3.43	4.25	4.38
		SD	0.70	0.49	0.44
		min	2.17	3.29	3.05
		max	4.89	5.33	5.16
		range	2.71	2.04	2.11
		n	39	39	39
cuts within 10 m	LAI	mean	3.11	3.76	3.94
		SD	0.51	0.48	0.55
		min	1.91	2.79	2.58
		max	4.70	5.11	5.29
		range	2.78	2.32	2.71
		n	104	104	104

Discussion

Initial Conditions

The univariate analysis shows that the Working Woods area has a mostly closed canopy in which initial canopy openness did not exceed 36%. Plots, when organized by treatment, had an average initial canopy openness between 17 and 21%, further indicating that this experimental area can be characterized as having a homogenously closed canopy. Even with low canopy openness values throughout the area, there did exist some variability in the initial canopy openness when examining future treatment influences. Specifically, trees that would not experience any release from competition via girdling had lower initial canopy openness than the trees that had some neighbors girdled in the winter of 2019 to 2020. As latitude and longitude had an effect on the initial canopy openness, it appears that the initial conditions are dependent on where the tree is located in Working Woods. This is valuable information for showing that this area has a mostly closed canopy, and indicates that, prior to management understory light availability in this forest was probably generally low.

Girdling's Influence on Canopy Structure

The canopy openness (derived from the hemispherical photos) appears to be, at most, only marginally influenced by girdling. The categorization of time since treatment cut and the relation the previous year's measures is what defines this complex trend (Figure 2.2, Table 2.2). As Figure 2.1 shows, the IC-0.5 treatment category appears to have a different canopy openness trend when compared against the control trees, as opposed to the other treatment categorizations. While this figure also suggest that LAI values, hen grouped by treatment category, tend to have little trend differences, but the

IC-0.5 category has a markedly lower trend. Both the categorization and the intensity of release show effects on LAI measurements made in 2020 (Table 2.5). Yet these observations are only seen in the two-factor models, and when spatial and environmental variables are included in a mixed effect linear model, the effects of girdling are not observed (Table 2.7). Similar observations are made with the DIFN measurements (Table 2.5). It appears that, whether a tree experiences cutting treatment or not is more important than the intensity of the cut area, for resulting in canopy structure changes. As the cut categories show in Figure 2.2, there might be a temporal factor involved in the extent to which this effect is realized as only the most recent treatment showed a trend differing from control trees.

When examining the subset of trees that had three years of canopy metric measurements, it appears that the canopy metrics have changed very little in response to the girdling treatment. Despite each metric showing a change in mean values after the treatment, these changes were observed in both trees that did receive treatment and those that did not. For each metric the trees that had girdling within a 10 meter radius had canopy structure metric values that indicated a more open canopy before treatment, compared to the trees that did not receive treatment. If the treatment resulted in increasing the openness of the canopy, then the trees receiving treatment should have seen a greater increase than the trees that did not receive treatment. Given that all the trees in these subsets exhibited similar increases in canopy openness and DIFN (and similar decreases in LAI), appears that the girdling treatment has not yet altered the canopy structure or its temporal variability in a substantial manner. Additionally, since the girdled trees take several years to fully die, we might expect a greater change in canopy structure after more time has passed. However, the greatest change in canopy structure occurred in the year immediately following treatment (Table 2.8). Yet due to the control trees exhibiting similar changes to the treated trees, this observation may not be a result of the girdling treatment itself.

Girdling Effects on Canopy Structure is not Dependent on Species or Initial Tree Size

In the mixed effect linear models neither the initial size nor the species had effects on any of the canopy structure metrics (Tables 2.4 & 2.5). This indicates that each of these species' canopies respond the same way when experiencing girdling treatment. Interestingly the initial size of the tree does not appear to influence the canopy structure in response to girdled treatment. As larger trees are known to have larger canopies, it might be expected that the larger individuals in this study would have lower canopy openness and would likely not see as much of a change in canopy structure in response to girdling treatment (Lusk et al., 2003). But, leaf area from neighboring trees also contributes to the canopy openness (or lack thereof) measured around the focal trees, likely obscuring the impact of the focal tree size on the canopy openness in its vicinity.

Future Implications

A potential explanation for the minor effects of girding on these canopy structure metrics could be that not enough competing trees were girdled. Other experiments have seen canopy structure changes as a result of girdling 65% of the canopy area or more (Grigri et al., 2020). With only 20% of the canopy experiencing any cutting, a larger amount of girdling may be required to observe significant canopy structure changes. Additionally, applying girdling in a less variable way would provide better context for the amount of girdling required to observe canopy structure changes. This study had a fairly heterogeneously application of the girdling area. A majority of the target trees that experienced a release from competition, had less than 3000 cm^2 basal area of girdled trees in a 10 m radius, yet the mean total basal area of cut trees within the 10 meter radius around the target tree was 3400.361 cm^2. A more homogenous spread of girdled area around the target trees might elucidate the effects of girdling intensity on the canopy structure. Additionally, given that there appears to be a trend of time since girdling treatment in effecting canopy structure, a wider range of girdling treatments that also take into account the intensity and time since cutting would be useful for further study. For example, eve for studies with more intense girdling treatments, it can take several years for an effect to be observed on canopy structure, and that effect can be very short-lived (Grigri et al., 2020).

Given the seemingly minor impacts that girdling treatment has on the canopy structure in this forest, continued observations are recommended. Girdled trees take ,on average four to five years to completely die (Fassnacht & Steele, 2016), and in this experiment by the 2020 canopy structure metrics, girdled trees have been girdled for at most, 1.5 years (the other group is only half a year post-treatment). While not quantified, observers have noted that many of the girdled trees in Working Woods are still leafing out as recently as the spring of 2021, 1.5 and 2.5 years after treatment. Continued measurements for additional years might allow for the treated trees to fully die, resulting in more observable treatment effects. In an experiment with a more intense girdling treatment in Michigan, the maximal effect of LAI was not observed until 3 years after treatment (Hardiman et al., 2013). Additionally, to completely understand how the girdling treatment effects the canopy structure, these methods for collecting canopy structure metrics around the target trees should be replicated around the girdled trees. Taking yearly measures including a measure prior to treatment would provide a understanding for how the girdling treatment influences the canopy of the girdled tree and therefore the understory light condition directly under the girdled tree. This would allow for not only verifying that these trees may take 4-5 years to die, but provide context for how fast the light conditions change underneath the girdled canopy.

As of the current measurements it appears that girdling results in very little changes to the overall canopy structure. This would indicate that the understory light conditions are not changing greatly either. This might indicate that if the forested area has a large invasive species population, that girdling may be a good option for removing unwanted trees, while not causing large changes in understory light availability that would allow for these invasive species to outcompete native understory plants.

The LAI measurements were found to have little reproducibility even when controlled for sky conditions (see methods section). Scatter correction application resulted in lower reproducibility as well. The cause of this is likely a result of the less standard LAI measurement protocol utilized in this study (López-Serrano et al., 2000). Future LAI measurements in Working Woods could be made along plot level transects to characterize LAI at the stand level (LI-COR, 2012), for comparison of treatment effects at the plot-level, as opposed to at the tree-level. This approach might increase the temporal reproducibility of the LAI measurements.

Even though automated hemispherical photos appeared to have less variable yearly measurements than LAI and DIFN (Table 2.2), this protocol is not without its flaws including, but not limited to, over exposure from the photos resulting in increased gap fraction and over correction when correcting for such exposure (Beckschäfer et al., 2013). An analysis using the manually processed hemispherical photos would be beneficial to corroborate the results from the automated processed photos. If manually processed photos produce similar results to the automated process, then arguably automated photos would be the better option for continued measurements. This could result in decreasing the labor for large sites like Working Woods and reduce the potential for observer bias in threshold application. Also, as my analyses showed, the subjectivity of manual processing of canopy photos can introduce error and bias when multiple people are involved in the processing of the canopy photos via manual methods.

Land Manager Recommendations

Considering that this treatment does not result in large changes to the canopy structure, this treatment would not be recommended to land managers aiming to increase canopy openness, or increase understory light availability. In a similar forest to Working Woods, an increase in canopy openness would further any invasive shrub that might already be present (Cunard & Lee, 2009; Taylor et al., 2017). If increasing light availability for an understory population of trees, or the promotion of late successional species, or more shade intolerant species would require greater amounts of girdling (Dale et al., 1995). Regardless of the reason for a land manager wishing to increase the amount or size of gaps in a forest, this treatment does not appear to result in the desired goal of increasing light availability in the forest, and will likely not bring along the benefits of increased gap size, at least no within the first two years after girdling.

In addition to land managers attempting to reach the specific goals of their forested area, mangers should consider how their forested area will respond to climate change projections. Secondary forests are likely to experience changes as a result of future climate change. Furthermore these changes are likely to be dependent on the ecosystem legacies as well as the diversity of the landscape (Frelich et al., 2020). Looking towards the future of their forested land will be essential for applying forest management techniques, and taking potential climate change into account is important for selecting a forest management technique despite the large amount of uncertainty related to climate change and forest management strategies (Keenan, 2015).

CHAPTER III

GIRDLING AS A TREATMENT FOR PROMOTING TREE GROWTH

Introduction

Secondary post-agricultural forests experience a variety of challenges due to a history of deforestation and replacement by agricultural land (Vellend, 2003). For example, in the northeastern U.S., where many forests are regrowing on agricultural land abandoned in the second half of the 20th century (Flinn & Marks, 2007) secondary forests are often comprised of dense stands of relatively slow-growing, even-aged trees represented by few species, especially 'generalist' species like red maple or shade tolerant species like sugar maple (Bilodeau-Gauthier et al., 2020; Poznanovic et al., 2013). Management of these secondary forests often aims for a variety of goals, including as well as adaptation to climate change (Flinn & Marks, 2007; Millar et al., 2007). Yet as these goals are pursued and as secondary forests regrow they are presented with additional challenges including a potential increase in invasive species that often coincides with the decrease of local biodiversity (Flinn & Marks, 2007; Meiners et al., 2008; Vellend, 2003). However, the effects these understory invasive species may have on overstory or midstory trees is less well known (Hartman & McCarthy, 2007). In addition to issues concerning invasive species management, optimal tree management strategies remain unclear and subject to specific management goals. Commonly used gap based methods can achieve goals of sustainable harvests and improve some ecosystems attributes (e.g. establishment of more shade-intolerant species) but, gap-based management also has some limitations (e.g. understory overgrowth, lack of coarse woody debris) (Kern et al., 2017). Other, less intense methods (those that keep a mostly continuous canopy) of forest management, such as single-tree selection cuts or girdling are often used if there is less emphasis on growth or regeneration of shade intolerant species (Vauhkonen & Packalen, 2019). In the northeastern U.S., common techniques to increase tree growth, regeneration, and/or diversity include group selection cuts to create relatively large canopy gaps (gap-based management), more localized selection cuts implemented via felling or girdling (e.g., targeting individual trees for removal and release from competition), and shelterwood cuts. Among these different techniques, girdling is perhaps the least well studied. The effects of girdling on the understory and residual trees, including potential crop trees, is less understood compared to more widely implemented and studied management strategies, like individual or group selection cuts (Lhotka et al., 2018).

Gap-based management techniques, like group-selection and single-tree selection, are the most commonly applied approaches to forest management in the Great Lakes region, but their adequacy is increasingly being questioned (Kern et al., 2017). Selection cuts are often implemented with the goal of promoting growth in remaining or future crop trees (Jones et al., 2009). Treatments such as group selection cutting, and shelterwoods cause large canopy gaps around target trees but the increased growth from these gaps does not necessarily appear until three to five years post-treatment (Jones & Thomas,

2004). The growth response of residual trees depends on the size and number of canopy gaps created and, the initial size of the crop trees (Bose et al., 2020). Larger residual trees see smaller responses to gap creation (Jones & Thomas, 2004) as do trees closer to the gaps (Dale et al., 1995). Smaller gaps and individual selection cuts generally favor increasing dominance of shade-tolerant species (Kern et al., 2012; Phillips & Shure, 2015; Webster & Lorimer, 2005). Larger gaps and group selection cuts are more likely to result in development of thick shrub layers, native or non-native, that can hinder regeneration early sapling growth (Kern et al., 2017).

Tree girdling, the process of cutting the phloem around the bole of a tree, is used to mimic natural disturbances like disease or pest infection and often results in a slow tree death; in a study of 756 individuals, 75-85% had died between 4 and 5 years of treatment (Fassnacht & Steele, 2016). This slow death of trees that are either unhealthy or undesirable (e.g., for lumber or diversity maintenance) results in the creation of snags (, or standing dead wood) which has significant benefits to forest ecosystems (Fassnacht & Steele, 2016). Girdling is commonly applied at the stand level to accelerate forest succession (Hardiman et al., 2013). However this increase in primary productivity was only observed in a similar study once girding reached 65% disturbance severity (Stuart-Haëntjens et al., 2015). Furthermore, this was observed only one year after girdling treatment was applied, indicating that many of the girdled trees likely need more time for greater release from competition ((Grigri et al., 2020). Other, similar efforts have focused on stand level carbon cycling after girdling disturbance found that stands that experience a less intense disturbance such as girdling did not have a decrease in gross primary productivity (Gough et al., 2013). However, girdling with the focus of releasing specific individuals from competition has not been as closely studied, despite the promise of girdling as an alternative to other treatments like group or individual selection cuts (which typically involve felling and removal of trees selected for cutting).

Understory vegetation is known to be influential in the regeneration of the seedling layer particularly when large gaps are created through clear cut or group selection harvests. Native and exotic shrub layer can offset the benefits of gap creation in forest stand (Kern et al., 2017). Such invasive species in the Northeast Ohio area including glossy buckthorn and multiflora rose are problematic in their immense spread in the area and have been shown to outcompete tree seedlings and saplings (Fagan & Peart, 2004; McDonald et al., 2008). However, very few studies have examined the effects of shrubs, particularly invasive shrubs, on trees larger than seedlings or saplings (Hartman & McCarthy, 2007). Thus, it is uncertain whether removal of invasive shrubs, such as often occurs as part of 'timber stand improvement', offers additional benefits to growth of crops trees released by various 'thinning strategies, like selection cuts or girdling.

Forest management's role in creating gaps has been found to have an impact on the biodiversity of forest environments. Shade tolerance differences between species has resulted in species composition changing as a result of this gap creation (Muscolo et al., 2014). The creation of gaps within the forest and as a result of forest management practices has led to increases in the relative abundance of shade tolerant species like sugar maples (Jenkins & Parker, 1998; Webster & Lorimer, 2005). Even the creation of small gaps through single cut selection favors shade tolerant species and thus results in less biodiversity (Neuendorff et al., 2007). It has also been found that not only the creation of gaps but the size of the gaps influences the success of more shade intolerant species. Larger gaps favor more mid-tolerant or shade intolerant species (Dale et al., 1995; Muscolo et al., 2014). However, there exists some evidence that shade tolerance is less of a predictor than pre-treatment conditions like overstory basal area or seedling density (Bose et al., 2020). In the case of species like *Liriodendron tulipifera* L. (tulip poplar), a shade intolerant species, it has shown to be successful in many different gap sizes due to its fast growing nature (Phillips & Shure, 2015). Additionally, tulip poplar is shown to have indeterminate growth which can allow it take advantage of lower light levels than other shade intolerant species (LeBlanc et al., 2020). Furthermore, tulip poplar seedlings have been found to respond greatly to intense release from competition as a result of group selection cuts (Williams, 1976). However, much of these studies were the result of single or group selection cuts and the canopy gaps caused as a result. The growth of different shade tolerant trees species after girdling treatment is less well understood.

The measure of tree growth can be performed in a variety of ways. While many studies of annual growth may make yearly DBH measurements, dendrometers allow for weekly measurements that can be used to access growth rates throughout the growing season (McMahon & Parker, 2015). Dendrometer bands have the additional benefit of allowing repleted measures of precisely the same location on the tree bole, allowing for more accurate estimates of annual diameter growth increment as compared to annual diameter measurements acquired with flexible measuring tapes (Cattelino et al., 1986). Dendrometer bands can vary in form and application with expensive automated options available as well as cheaper steel bands that require manual measurements. Furthermore these cheaper options are found to be closely related to the more expensive models (Just & Frank, 2019). These can be applied over a large area and to a large amount of trees for minimal labor costs.

Despite all the work to examine how forest systems are influenced by specific management processes, very few studies have focused on how midstory trees may be influenced by these treatments (Jones & Thomas, 2004). Even less has been studied on the effect of girdling on midstory trees (Grigri et al., 2020) and the possible effects of timber stand improvement on the growth of trees released by low intensity 'thinning' treatments like selective girdling or felling of individual trees. This study aims to address these gaps in knowledge. Before and after implementing a girdling treatment, with and without additional timber stand improvement to remove grapevines and invasive shrubs, we measured the growth of 457 trees, 296 via seasonal measurement dendrometer bands and 161 via annual dbh measurements. Growth was monitored for up to three years after forest management treatments were implemented. We hypothesize that targeted girdling treatments, which mimicked single tree selection cuts, will result in increased growth rates of midstory trees that experience this slow release from competition. Furthermore, by addressing the amount of release from competition around each target midstory tree this study aims to show that growth is dependent on the amount of release from competition due to girdling. Additionally, we expect the growth effects of the girdling treatment will be largest for the least shade tolerant species (tulip poplar) in our sample population, smallest for the most shade tolerant species (sugar maple) in our population, and intermediate (red maple). As with shade tolerance, the initial size of the midstory tree is likely to influence the growth effects as a result of girdling treatment; specifically, we expect lager positive growth responses to girdling for smaller trees that presumably were initially more suppressed by competition with neighbors. Lastly, following Hartmann and McCarthy (2007), we expect that removal of invasive shrubs as part of timber stand improvement will result in additional benefits to growth rate of crop trees, especially for smaller trees.

Methods

Site Description and Experimental Design

This site was located within the area of the Working Woods at Holden Arboretum in northeast Ohio. This site was previously agriculture land that has been abandoned for around 60 years. The area is young forest that is mostly dominated by red maple and tulip poplar. This area was divided into nine 100 meter x 100 meter plots that were further divided in half to apply treatments in separate years. A full description of the plot treatments and application of these treatments can be found in the methods of Chapter II. A total of 457 trees were identified and tagged to measure a variety of metrics throughout the Working Woods area. Figure 2.1 shows the site and location of each target tree. Trees were selected that were midstory or understory trees that experienced a large amount of competition from neighboring trees.

Treatments were applied either during the winter of 2018 or the winter of 2019. Trees were girdled that were in poor health or form (e.g. multiple stems), were undesirable species, or individuals that were too close to target trees. Most of the trees that were selected for removal were girdled, but those that were too close to target trees or public trails were felled and left on the forest floor in place. Table 2.1 describes the population of trees that received 'cutting' (girdling or felled) treatment.

GIS

The latitude, longitude, and elevation of each target tree and each girdled or felled tree was recorded using a Garmin GLO 2 GPS receiver. These locations were recorded during the winter of 2019-2020, which allowed for improved signal strength and accuracy due the senescence of deciduous leaves. QGIS was used to identify 10 m radii around each target tree (QGIS Development Team, 2021). The number, species, and size of each girdled or felled tree within each 10 meter radius was used for determining release from competition.

Growth Measurements

To measure the growth of the selected target trees, two methods were used depending on the size of the tree. For most trees (n = 269), an initial, or pre-treatment dbh measurement was observed in May or early June of 2018. Larger trees (> 5 cm dbh) were typically fitted with stainless steel dendrometer bands to monitor growth more frequently and precisely. The expandable 'measurement window' of the bands was measured weekly or biweekly from April until the end of September from 2018 to 2020. Measurements were made weekly in the beginning and end of the growing season to establish when growth started and slowed/stopped. Digital calipers were used to measure the width of expandable window of dendrometer bands (OEM 25363 Digital Caliper). The dbh of these banded trees was also recorded at the beginning of the growing season in 2019 and 2020. Three species were focused on for these large banded trees; *Acer rubrum* L. (red maple), *Acer saccharum* Marsh. (sugar maple), and *Liriodendron tulipifera* L. (tulip poplar). The initial dbh of the large target trees was recorded at the beginning of the 2018 growing season, with the tulip poplars showing the largest average initial size, and sugar maple showing the smallest average initial size (Table 3.1). For smaller trees (< 5 to 10 cm dbh) that were not fit with dendrometer bands, dbh was measured at the beginning of the growing season in 2018 and 2019 and at the end of the growing season in 2020 (in November). These measurements will be designated as either the dendrometer measurements or larger (L) trees and the smaller trees which received only yearly dbh measures (S). The population of smaller monitored trees is composed mostly of three species, *Acer rubrum* L. (red maple), *Acer saccharum* Marsh. (sugar maple), and members of the *Ulmus* genus (elms). The initial size of these species was taken at the beginning of the growing season in 2018 (Table 3.2).

To differentiate between dbh measures, measurements were denoted as either initial dbh of the specific year (dbh^i) or the final dbh of the specific year (dbh^f). Final yearly dbh measures for the large, banded trees was calculated by adding the measurement window increment to the initial dbh. This increase was added to the circumference of the tree taken from the initial dbh measurement. This final circumference measure was then converted back into dbh for the final dbh of the year. Growth measurements for small trees was found through the difference between yearly initial measurements. Growth values were established for three growing periods, the first being growth from 2018 to 2019, which was found from the difference between initial 2019 and 2018 dbh measures for all trees. The second being growth from 2019 to 2020, which was the difference between the initial dbh measurement of 2019, and depending on the size of the tree, the final dbh measure of small, non-banded trees measured in 2020 or for the banded trees, the measurement window increment added to the initial dbh measurement in 2020. The third being the growth from 2018 to the end of 2020, which was the difference between the initial 2018 dbh measurements and the final dbh measure of small, non-banded trees measured in 2020 and, for the banded trees, the measurement window increment added to the initial dbh measurement in 2020.

TABLE 3.1 Banded trees initial size.
Fifteen trees that had died or been severely damaged during the experiment were not included. Eleven red oak and one tree that was located outside of the plots were not included. Diameter at breast height is represented in cm.

	red maple			sugar maple			tulip poplar			Total		
Plot Treatment	n	Initial Size (dbh)	Standard Deviation	n	Initial Size (dbh)	Standard Deviation	n	Initial Size (dbh)	Standard Deviation	n	Initial Size (dbh)	Standard Deviation
C	16	17.3	5.86	17	17.63	6.83	14	33.80	12.28	47	22.77	11.5
IC	48	16.20	6.35	43	12.62	5.06	49	26.17	9.32	140	18.32	9.13
IC+TSI	34	19.77	6.25	20	12.39	4.10	28	33.00	7.69	81	22.03	10.23
Total	98	17.70	6.40	80	13.50	5.51	91	29.53	9.91	269	20.23	10.05

TABLE 3.2 Unbanded Trees Initial Size
Three trees that had died or been severely damaged during the experiment were not included. Three trees that were located outside of the plots were not included. Diameter at breast height is represented in cm.

	red maple			sugar maple			elm			Total		
Plot Treatment	n	Initial Size (dbh)	Standard Deviation	n	Initial Size (dbh)	Standard Deviation	n	Initial Size (dbh)	Standard Deviation	n	Initial Size (dbh)	Standard Deviation
C	13	5.47	1.74	20	3.92	6.83	6	5.23	2.42	39	4.7	1.86
IC	8	5.35	1.23	30	4.52	1.40	21	5.18	1.54	59	4.87	1.45
IC+TSI	12	7.33	1.59	12	4.81	1.48	13	6.17	1.92	55	5.68	1.89
Total	33	6.17	1.78	80	4.52	1.46	40	5.51	1.82	153	5.14	1.76

Neighborhood Competition Indices

In order to assess the competition around each target tree several neighborhood competition indices (NCI) were developed. The models were based off Hartmann and Messier's (2011) models which factored in size of the competitor, and distance of target tree to the competitor. The general model for such neighborhood competition indices is shown below. NCI models were used to assess both the amount of release target trees were experiencing as a result of girdled treatment, as well as the amount of living competition each tree experienced. Both girdled and living NCIs were assessed for a 10 meter radius around each target tree. Species specific NCI models were also made for the three main target tree species (sugar maple, red maple, and tulip poplar) as well as a red oak model. Other species were designated as other. This categorization allowed for intra and interspecific models for each target tree. Variable description of this model can be found in the methods section of chapter II.

$$NCI_i = \sum_{j=1}^{N} \frac{(DBH_j)^{\alpha}}{(dist_{ij})^{\beta}}$$

To determine the best NCI models were assessed through subset regression using the growth from 2018 to 2020 as the dependent variable. The initial size or dbh measure in 2018 was included as a covariate as it is highly correlated with the growth from 2018 to 2020. The NCI models were compared in subset regression using the *regsubsets* function from the leaps package for R software (Miller & Lumley, 2020; R Core Team, 2019). Subset regression models were made for the species specific NCI models resulting in three initial models including *i)* total cut area, *ii)* intraspecific cut area, and *iii)* interspecific cut area. In each of these models all four different β values were tested. The best NCI value and any others that were within 2 ΔBIC of the lowest BIC were included in the final NCI model. From this final model the five lowest BIC values included intraspecific cut area β0.5 (-27.827), intraspecific cut area β0 (-27.727), intraspecific cut area β1 (-27.409), and total cut area β0 (-25.577). Due to the lower sample size of the intraspecific cut area (n=140), versus the higher sample size of the total cut area (n=318), as well as the low ΔBIC (2.25), both the intraspecific cut area and the total cut area NCIs with a β of 0 were included in the univariate analysis.

To ensure that trees receiving girdling treatment within the 10 meter radius were not more likely to have large release of competition due to already having greater competition around them, the relationship between living basal area and cut basal area was observed. Figure 3.1 shows the relationship between these two variables and shows that there is no indication that the trees receiving more release from competition have more living competition to begin with.

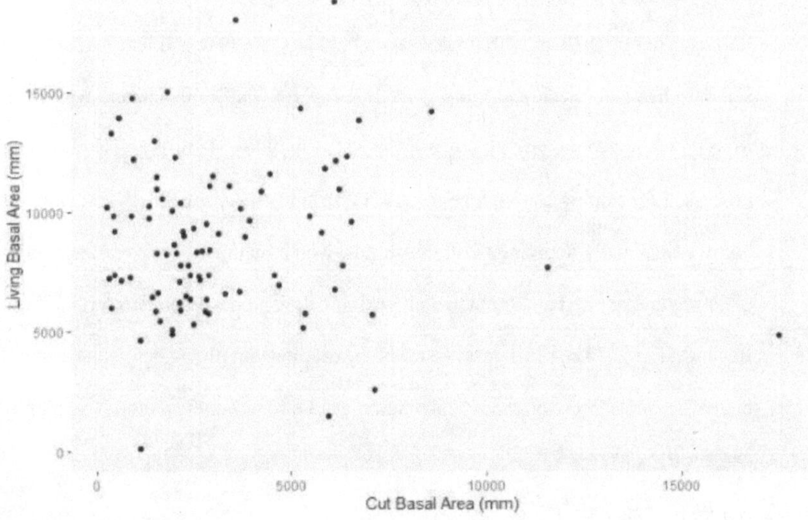

FIGURE 3 1 Relationship between cut area and living area.

Any tree without girdling in the 10 meter radius was not included (n=91).

LAI 2200 Measurements

LAI values were collected using a LI-COR 2200C Canopy Analyzer for 356 target trees in 2019 and 388 target trees in 2020 (LI-COR, 2012). LAI measurements were taken in each cardinal direction around a target tree similar to López-Serrano et al., method (2000). A full description of the collection and calculation of the LAI measurements can be found in the methods section of chapter II.

Hemispherical Photography

Hemispherical photos were taken in 2018, 2019 and 2020 at 122, 447, and 349 target trees respectively within the study location. As for LAI reading, variability in the number of trees measured each year reflects variability in the time and labor available for measurements each year. Photos were taken in 2018 between late July and late August. Photos taken in 2019 and 2020 were taken between late June and late August. Following standard hemispherical photography methods, a NIKON D5600 camera was attached to a tripod to sit above the ground cover foliage (Rich, 1990). Hemispherical photos were processed for finding canopy openness (CO) using two separate methods. One method involved manually applying pixel thresholds, which includes observer bias, while the other method applied an automated algorithmic threshold to each photo (Glatthorn & Beckschäfer, 2014). A full description of the methods involving hemispherical photo collection and canopy openness calculation can be found in the methods section of chapter II.

Health Assessment

In 2021, a health assessment was performed for a subset of the target trees, focusing on the larger trees (n=286). This health assessment featured a canopy assessment as well as tree health/damage and site characteristics around each target tree. Canopy measures included canopy position canopy light exposure, and crown class. These were measured following standard forestry practices (Bechtold, 2003). Additionally included in the canopy assessment was the number of live and girdled overtopping trees at each target tree canopy. Trees were counted as 'overtopping' if they had branches that reached over any part of the canopy of the target tree. Health-related metrics that were measured included crown, root flare lower trunk damage, other damage to the trunk, bare soil ground cover, microtopography assessment, and overall tree condition. These indices of tree health were measured following the protocols established by Vogt et al (2014). Full descriptions of each metric can be found in the methods section of chapter II.

Surveys of Invasive Shrubs

Focusing on a 4 m radius around most of the target trees (n=255), surveys were done to assess ground cover of invasive and native shrubs. Cover was measured in each year, with 2018 focusing only on invasive species. As such, only invasive species cover is reported for each year. However, native shrubs are relatively rare and when they were presented account for less than 20% of total shrub cover on average. Invasive shrub cover is reported as the midpoint of one of the six cover classes recorded by the observers (class 0 = 0% cover, class 1 = 2.5% cover, class 2 = 15% cover, class 3 = 37.5% cover, class 4 = 62.5% cover, class 5 = 85% cover, class 6 = 97.5% cover). Invasive shrub cover was averaged over the growth periods to provide an average cover for that growth period (growth periods were defined as pretreatment growth from 2018 to 2019, growth from 2019 to 2020 and growth from 2018 to 2020).

Soil Texture Analysis

For a subset of the target trees (n=118), soil cores were taken in winter 2019 approximately 2 m from the base of the tree, and in each cardinal direction. Soil cores were divided into 0-20 cm deep sections and 20-40 cm deep sections. Plant roots and other organic material was removed from these soil core sections and the soil was sieved to separate into two fractions, that greater than and less than 2 mm in diameter. The fraction less than 2 mm was then used to measure the percentage of clay, silt, and sand particles using the hydrometer method from Gee and Bauder (1986). Soil samples were left to air dry and from this a bulk density was also found for each sample section (Arya

& Paris, 1981). For both bulk density and each soil texture composition value, the values between the two sections were averaged over both section samples, providing a 0-40 cm value for each.

Statistical Analyses

Univariate analysis. Simple (univariate) linear models were used to assess the effects of a variety of ecological variables on each of the three growth periods and the initial, pre-treatment conditions to describe the correlations among the different periods of growth, and correlations that existed prior to treatment application. Variables assess in this manner included potential spatial covariates (latitude, longitude, elevation at each target tree), the various metrics of tree health and the canopy position, total and intraspecific cut area and the living area in a 10 meter radii around each target tree.

Several different variables were used to describe whether, and to what extent, each targe tree 'experienced' the IC treatment. First, each target tree was separated into a category, which designated whether a target tree had experienced cutting within their 10 meter radius, and how many years had passed, since the cutting was implemented. Category designation included "C" for trees that did not experience any cutting, "IC-0.5" for trees that had cutting the winter prior that year's measures of canopy structure, "IC-1.5" for trees that had cutting two winters prior to that year's measures of canopy structure and "IC-0.5+1.5" for trees that had cutting occur both one and two winters prior to that year's measure of growth and size. The effects of NCI from trees that 'experienced' a release from competition (i.e. any trees without cutting in a 10 meter radius were not included).

The effects of each growth variable was tested in separate linear models against each dependent variable and spatial covariates. The lm function in the R software (version 1.2.1335) was used to create these statistical models and corresponding model statistics.

Three-factor model. For each growth period, treatment effects were assessed in a model that included the initial size, the species of the target tree and the cutting category, or the cut intensity, as well as interaction terms between the treatment and initial size, and the treatment and species. These models allow for effects of treatment to be interpreted as an increase in growth amount to be a result of the treatment. This model was used due to the large variation in initial sizes of trees within the plots, and different average initial sizes of the species. Three-factor models were specified with the lm function, and summarized with the Anova function from the car package in the R software (Fox & Weisberg, 2019).

Multivariate least squares models. A subset of potential predictors including variables from the univariate analysis were assessed in multivariate least squares models. Predictors were included in these models if: *i)* the results of the simple ANOVA or regression models showed evidence for influencing the growth, *ii)* the predictors were useful for assessing how these growth periods might be similar along a spatial gradient (e.g. latitude, longitude, elevation of each target tree), or *iii)* if they were important to testing original hypotheses (e.g. species, invasive shrub cover). These multivariate least squares models were specified using the lm function and the effects of the predictors were assessed using Type II ANOVA the Anova function from the car package for the R software (Fox & Weisberg, 2019).

Temporal changes in growth. To assess how growth rates may have changed over time as a result of the treatment application, basal area increment (BAI) and diameter increment over three years was assessed. The years were -0.5 prior to treatment, 0.5 years since treatment, and 1.5 years since treatment. Due to this assessment, only the trees in 2018 plots were used (total n=128). Trees were categorized as either IC, for trees that had girdling within a 10 meters (n=99), or C for trees that did not have girdling within a 10 meter radius (n=29). All of the trees selected were L trees that had been fitted with dendrometer bands. Mean values for dendrometer increment and BAI were used. Standard deviation of these two growth metrics were also calculated to provide a measure of how variable these growth metrics were.

Results

Univariate Analysis

Pretreatment variability. Initial dbh in 2018 was included as a dependent variable to assess if any spatial or environmental variables influenced the initial tree size. From these initial values, species had significantly different initial sizes between them with tulip poplar being the largest average tree (mean dbh =29.2 cm), followed by red maple (mean dbh = 14.6 cm), sugar maple (mean dbh = 9.01 cm) and elm (mean dbh = 5.51 cm) being the smallest. Each of the canopy metrics including, crown position (R^2 = -0.44, p<0.001), crown light exposure (R^2 = 0.36, p<0.001), live overtopping (R^2 = -0.47, p<0.001), dead/girdle overtopping (R^2 = -0.16, p<0.001), and crown dieback (R^2 = -0.034, p<0.01), all acted as significant predictors for the initial size of the tree. Several health metrics, including root flare (R^2 = -0.45, p<0.001), other damage (R^2 = -0.018, p<0.01), bare soil ground cover ($R^2 = -5.54 \times 10^{-5}$, p<0.001), and microtopography category (R^2 = -0.012, $p<0.05$) all were significant predictors to initial tree size in 2018. Soil moisture had an effect on the initial size of the tree (R^2 = 0.029, $p<0.001$). Canopy openness in 2018 (R^2 = 0.013, $p<0.05$), and overall tree condition (R^2 = -0.037, $p<0.001$), particularly those in good condition were also related to initial size.

Covariance of growth occurring from 2018 to 2019 with other variables. Both longitude and latitude were strong predictors for this growth period (R^2 = -0.063, $p<0.001$). The initial size (R^2 = 0.21, $p<0.001$) and species (R^2 = 0.18, $p<0.001$) were strong predictors for this growth period. All of the crown metrics, including crown position (R^2 = -0.20), crown light exposure (R^2 = 0.17), canopy class (R^2 = -0.23), live overtopping (R^2 = -0.13), and dead/girdle overtopping (R^2 = -0.048) were all significant predictors for this growth period ($p<0.001$). Microtopography (R^2 = 0.014, $p<0.05$) and overall tree condition (R^2 = 0.030, $p<0.01$) were the only health assessment variables that were significant predictors for this growth period. The invasive cover in 2018 and 2019 were not significant predictors for this growth period model. Neither the cut category nor the cut intensity were significant predictors.

In the models examining the growth from 2018 to 2019, the cut categorization of 2019 and the cut intensity measure of cuts occurring in the winter of 2018-2019 were used. This meant that the cuts that were made in the winter of 2019-2020 were not included.

Covariance of growth occurring from 2019 to 2020 with other variables. In these models the growth occurring from 2019 to 2020 was assessed using the cut categorization from 2020 and the cut intensity including both cuts made in the winter of 2018-2019 and the winter of 2019-2020. Unlike the previous model, the latitude and longitude was not a significant predictor for this year's growth (Table 3.3). However, the initial size (R^2 = 0.33, $p<0.001$), species (R^2 = 0.30, $p<0.001$), were both positive correlated with this growth period. The canopy metrics, except for crown light exposure, which was positively correlated, were all minorly negatively correlated for this year's growth ($p<0.001$). The overall tree condition was a significant predictor ($p<0.001$), while the microtopography was not. In this model, the cut intensity was a significant predictor the growth occurring in period ($p<0.05$). The variables that explained the most variability in growth from 2019 to 2020 were initial size (R^2 = 0.33, $p<0.001$) and species (R^2 = 0.30, $p<0.001$).

Covariance of growth occurring from 2018 to 2020 with other variables. In these models the growth occurring from 2018 to 2020 was assessed using the cut categorization from 2020 and the cut intensity including both cuts made in the winter of 2018-2019 and the winter of 2019-2020. Both longitude (R^2 = -0.028) and latitude (R^2 = -0.027) were strong predictors and negatively correlated with this growth period ($p<0.01$). Greater canopy openness results in a decrease in growth throughout this growth period (R^2 = -0.013, $p<0.05$). Like the previous model, the initial size (R^2 = 0.34), and species (R^2 = 0.38) were both significant predictors for this year's growth ($p<0.001$). The canopy metrics had mixed effects on the growth during this time period with only one (crown light exposure), having a positive effect and the rest (crown position, canopy class, live overtopping, dead/girdle overtopping, and crown dieback) all having negative effects (Table 3.3). The overall tree condition was a significant predictor (R^2 = 0.081, $p<0.001$), as was the cut intensity (R^2 = 0.020, $p<0.05$).

The intraspecific cut area was not a significant predictor (R^2 <0.017) in any of the models. Additionally, the cut categories (R^2 <0.017) were not significant predictors in any of the models. None of the soil texture variables (R^2 <0.04) nor the elevation were important to either the initial conditions or any of the growth periods. LAI and DIFN measurements made in both 2019 and 2020 were not significant to any of the dependent variables (R^2 <0.01).

TABLE 3.3 Covariate and variable assessment on dependent growth variables. R^2 values are reported, and significant p values are denoted with asterisks. *** < 0.001, **<0.01. *<0.05.

Variable	2018 dbh^{l}	2019dbh^{l}-2018dbh^{l}	2020dbh^{l}-2019dbh^{l}	2020dbh^{l}-2018dbh^{l}
Latitude	-0.0071	-0.063 ***	0.0014	-0.027 **
Longitude	-0.0019	-0.082 ***	0.0031	-0.028 **
Elevation	5.496×10^{-5}	-0.00019	0.0015	4.195×10^{-6}
% Sand 0-20 cm	-0.00043	-0.035	-0.0077	-0.033
% Sand 0-40 cm	0.0014	-0.046	-0.0025	-0.033
% Clay 0-20 cm	0.0044	0.028	0.0041	0.020
% Clay 0-40 cm	0.016	0.040	0.0036	0.032
Soil Moisture (% VMC)	0.029 ***	0.022 **	0.0023	0.00087
% Canopy Openness 2018	0.0127 *	0.0044	-0.00011	0.0011
% Canopy Openness 2019		0.0071	.-9.728×10^{-5}	0.0022
% Canopy Openness 2020			-0.0023	-0.013 *
% Invasive Shrub Cover 2018	0.014	-0.010	0.0092	0.00039
% Invasive Shrub Cover 2019		0.00066	0.0027	0.0010
% Invasive Shrub Cover 2020			-5.053×10^{-8}	-0.00091
Initial Size (dbh2018)		0.21 ***	0.33 ***	0.41 ***
Species	0.50 *** (tp>rm>sm>elm)	0.18 *** (tp>rm>elm>sm)	0.30 *** (tp>rm>elm>sm)	0.38 *** (tp>rm>elm>sm)
Crown Position	-0.44 ***	-0.20 ***	-0.23 ***	-0.32 ***
Crown Light Exposure	0.36 ***	0.17 ***	0.24 ***	0.32***
Canopy Class	0.53 ***	-0.23 ***	-0.30 ***	-0.41 ***
Live Overtopping	-0.47 ***	-0.13 ***	-0.23 ***	-0.28 ***
Dead/Girdling Overtopping	-0.16 ***	-0.048 ***	-0.036 ***	-0.064 ***
Crown Die Back	-0.034 **	-0.058	-0.034 ***	-0.027 **
Root Flare	-0.45 ***	-0.047 ***	-0.097 ***	-0.11 ***
Lower Trunk Damage	0.0069	-0.00013	0.00072	0.00049
Other Damage	-0.018 **	-0.010	-0.0090	-0.012 *
Bare Soil Ground Cover	-0.055***	-0.0039	-0.0067	-0.0066
Microtopography Category	-0.012 *	0.014 *	-0.011 *	-0.00077
Overall Tree Condition	-0.037 ***	0.030 **	0.0062 ***	0.063 ***
LAI 2019		0.0035	0.0062	0.012
LAI 2020			1.89×10^{-5}	7.86×10^{-5}
DIFN 2019		0.0015	-0.0030	0.00058
DIFN 2020			0.0062	0.0052
Living Basal Area	0.0022	0.0080	-0.015	-0.025
Cut Basal Area		-0.0026	0.0099 *	0.012 *
Cut Categorization		0.0068	0.017	0.021
Cut Intraspecific Basal Area		.017	0.0025	.0041

Three-Factor Analysis

In these models, each growth period was assessed against the cut category or intensity, the initial size, and the species identity, plus the interactions between the treatment and both initial size and species identity. Each model was tested for either the cut category or the cut intensity. In each of these models the treatment alone did not have a significant effect on growth (p>0.12198, Table 3.4). Initial size of the tree was a significant predictor for all models regardless of which growth period was examined and whether treatment was categorized or measured as the intensity of treatment (p<0.001). Species was also important for every model except for the model containing growth from 2018 to 2019 and containing cut intensity (p>0.05). The only interaction term that acted as a significant predictor was the interaction between species and cut intensity for model examining the growth occurring from 2018 to 2020 (p <0.05).

TABLE 3.4 Two-factor models with p values reported for each growth period Interaction terms were noted in the variable column with an asterisk between the two variable. Significant p values are denoted with asterisks. *** < 0.001, **<0.01. *<0.05.

Variables	Growth from 2018 to 2019		Growth from 2019 to 2020		Growth from 2018 to 2020	
	cut category	cut intensity	cut category	cut intensity	cut category	cut intensity
Cut category/intensity	0.12	0.38	0.88	0.53	0.16	0.20
initial size	<0.001 ***	<0.001 ***	<0.001 ***	<0.001 ***	<0.001 ***	<0.001 ***
Species	0.0046 ***	0.08935	<0.001 ***	<0.001 ***	<0.001 ***	<0.001 ***
initial size * cut category/intensity	0.10	0.69	0.49	0.33	0.71	0.41
species* cut category/intensity	0.63	0.78	0.96	.0.12	0.93	0.027 *
n	365	160	365	281	369	283

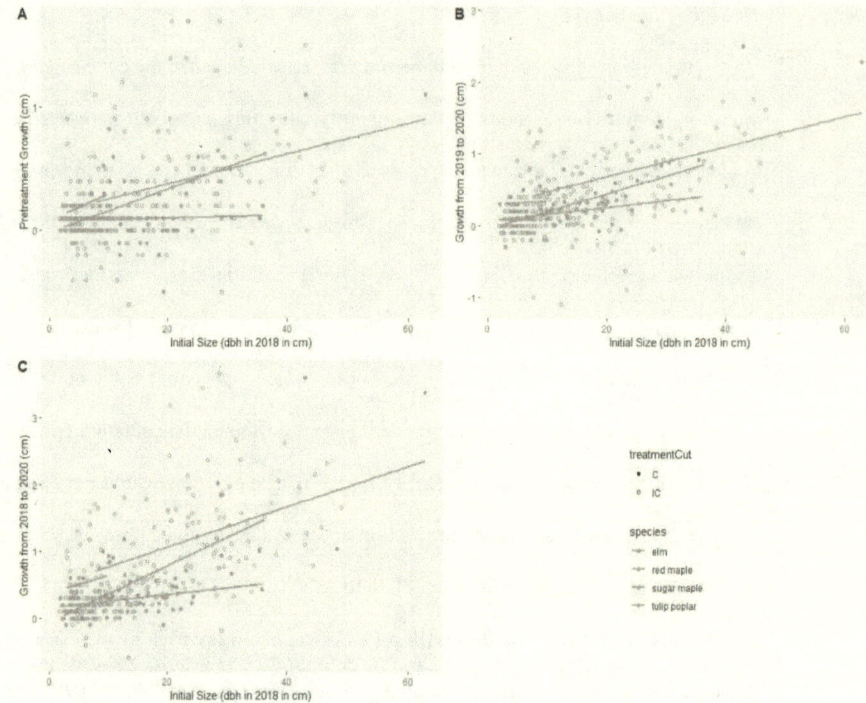

FIGURE 3.2 Relationship between growth and initial size

The relationship between the initial size and A) pretreatment growth, B) growth from 2019 to 2020, C) growth from 2018 to 2020. Species are shown by different colored variables and whether or not the tree experienced any cutting in the 10 m radius was shown by an unfilled in circle.

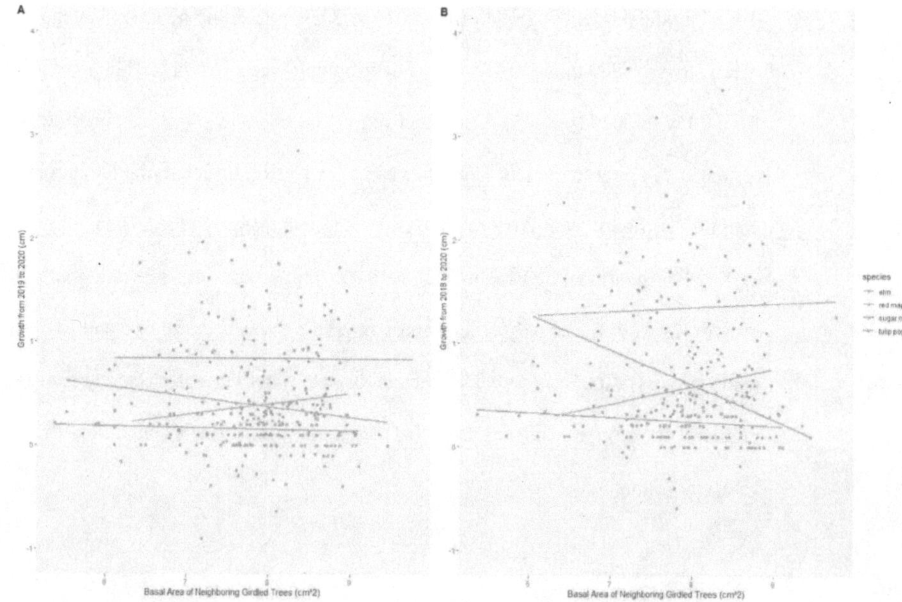

FIGURE 3.3 Relationship of growth and girdled area

The relation between the girdled area (intensity) and A) growth from 2019 to 2020, and B) growth from 2018 to 2020. Both graphs show a log transformed basal area of neighboring girdled trees. Colors of points and lines show the different species values.

Multivariate Least Squares Model

The multivariate least squares models were run with either the cut category (Table 3.5) or the cut intensity present in the model (Table 3.6). The models containing cut category did not show cut category on its own to be a strong predictor. It was a strong predictor in interaction terms with soil moisture for the growth from 2018 to 2019 ($p<0.05$). Soil moisture when not in an interaction term was a significant predictor as well ($p<0.001$). Species was the only other variable that acted as a significant predictor for these pretreatment growth models ($p<0.05$).

In the model for growth from 2019 to 2020 containing treatment category, species was a strong predictor ($p<0.001$). Live overtopping was a strong predictor for growth from 2019 to 2020 ($p<0.05$). The initial size of the tree was significant predictors ($p<0.05$). No interaction terms or other variables were significant predictors for the model containing growth from 2019 to 2020 and treatment category (Table 3.5).

For the model of growth from 2018 to 2020, species was a strong predictor for growth ($p<0.001$). The only other important predictors was initial size and live overtopping ($p<0.01$ and $p<0.05$ respectively). No other variables acted as a strong predictor for growth occurring from 2018 to 2020 in the models containing the treatment category (Table 3.5).

TABLE 3.5 Multivariate least squares models of growth and cut categorization P values are reported, and significant p values are denoted with asterisks. *** < 0.001, **<0.01. *<0.05.

Variables	Growth from 2018 to 2019	Growth from 2019 to 2020	Growth from 2018 to 2020
fixed effects			
treatment(category)	0.80	0.79	0.88
Longitude	0.052	0.81	0.22
Latitude	0.56	0.41	0.75
Elevation	0.71	0.84	0.86
Initial Size	0.033	0.028 *	0.0036 **
Average Invasive Shrub Cover	0.073	0.65	0.36
Soil Moisture	<0.001***	0.078	0.59
Live Overtopping	0.49	0.014 *	0.018 *
Dead/Girdle Overtopping	0.19	0.34	0.063
Overall Tree Condition	0.81	0.25	0.55
LAI	0.97	0.55	0.76
DIFN	0.46	0.091	0.29
Species	0.038*	<0.001***	<0.001 ***
Interaction Effects			
treatment(category) * initial Size	0.38	0.067	0.58
treatment(category)*species	0.66	0.99	0.99
treatment(category)*shrub cover	0.14	0.15	0.25
treatment(category)*live overtopping	0.60	0.22	0.59
treatment(category)*dead/girdle overtopping	0.67	0.96	0.99
treatment(category)*crown die back	0.47	0.96	0.96
treatment(category)*overall tree condition	0.17	0.99	0.92
treatment(category)*soil moisture	0.023*	0.32	0.93
treatment(category)*LAI	0.26	0.73	0.79
treatment(category)*DIFN	0.061	0.57	0.42
n	248	311	315

The cut area intensity was not a significant predictor for any of the models shown in Table 3.6. However, for growth occurring from 2018 to 2019, it was a significant predictor when included in an interaction term with soil moisture ($p<0.05$). Soil moisture was a strong predictor for this growth period ($p<0.001$). Initial size and species also acted as important predictors for pretreatment growth for the models looking at cut intensity ($p<0.05$).

In the models of growth from 2019 to 2020, species was significant for this model ($p<0.001$, Table 3.6). In this model, initial size was also a significant predictor ($p<0.05$). Live overtopping also acted as a significant predictor for this growth period ($p<0.05$). Species was a strong predictor and was a stronger predictor than the pretreatment growth ($p<0.001$). None of the interaction terms acted as significant predictors in this model (Table 3.6).

In the model of growth from 2018 to 2020, species was again a significant predictors ($p<0.001$). Like the 2019 to 2020 model, initial size and live overtopping both were significant predictors ($p<0.01$ and $p<0.05$ respectively). None of the interaction terms acted as significant predictors and no other variables acted as significant predictors on their own (Table 3.6).

TABLE 3.6 Multivariate least squares models of growth and cut intensity. P values are reported, and significant p values are denoted with asterisks. *** < 0.001, **<0.01. *<0.05.

Variables	Growth from 2018 to 2019	Growth from 2019 to 2020	Growth from 2018 to 2020
Fixed Effects			
treatment(intensity)	0.80	0.79	0.88
Longitude	0.052	0.81	0.22
Latitude	0.56	0.41	0.75
Elevation	0.71	0.84	0.86
Initial Size	0.033*	0.028*	0.0036**
Average Invasive Shrub Cover	0.073	0.65	0.36
Soil Moisture	<0.001***	0.078	0.59
Live Overtopping	0.49	0.014*	0.018*
Dead/Girdle Overtopping	0.19	0.34	0.063
Overall Tree Condition	0.81	0.25	0.55
LAI	0.97	0.55	0.76
DIFN	0.46	0.091	0.29
Species	0.038*	<0.001***	<0.001***
Interaction Effects			
treatment(intensity) * Initial Size	0.38	0.067	0.58
treatment(intensity)*species	0.66	0.99	0.99
treatment(intensity)*shrub cover	0.14	0.15	0.25
treatment(intensity)*live overtopping	0.60	0.22	0.59
treatment(intensity)*dead/girdle overtopping	0.67	0.96	0.99
treatment(intensity)*crown die back	0.47	0.96	0.96
treatment(intensity)*overall tree condition	0.17	0.99	0.92
treatment(intensity)*soil moisture	0.023*	0.32	0.93
treatment(intensity)*LAI	0.26	0.73	0.79
treatment(intensity)*DIFN	0.061	0.57	0.42
n	248	311	315

Temporal Changes in Growth

Before girdling treatment were implemented, the IC trees had higher mean and median growth than control trees that did not have any girdled trees within a 10 meter radius (Figure 3.4; 3.5; Table 3.7). However, the distribution of growth rates was substantially overlapping for trees that did and did not have girdling nearby, and the distribution of growth rates among trees was also similar across years, from pre-treatment to approximately one and a half years after girdling (Figure 3.4 & 3.5). Thus, no strong temporal or treatment effects on growth were apparent when focusing on this subset of trees.

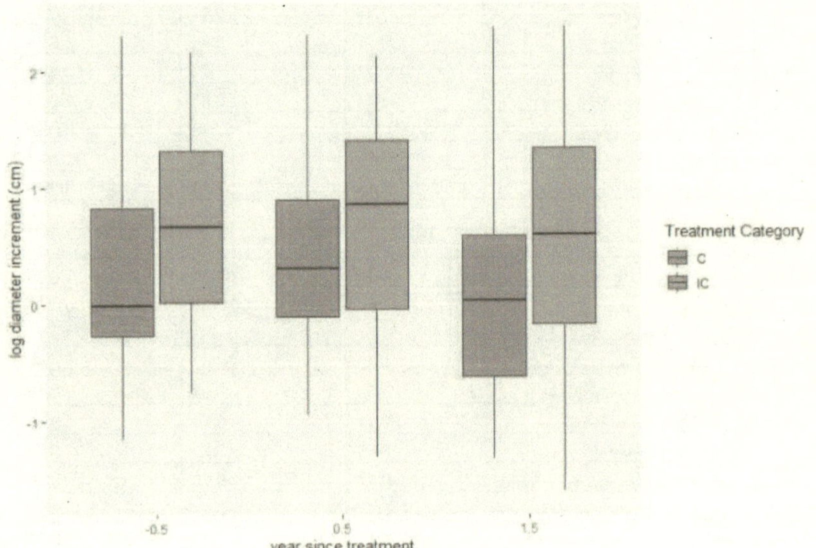

FIGURE 3.4 Diameter increments over time since treatment

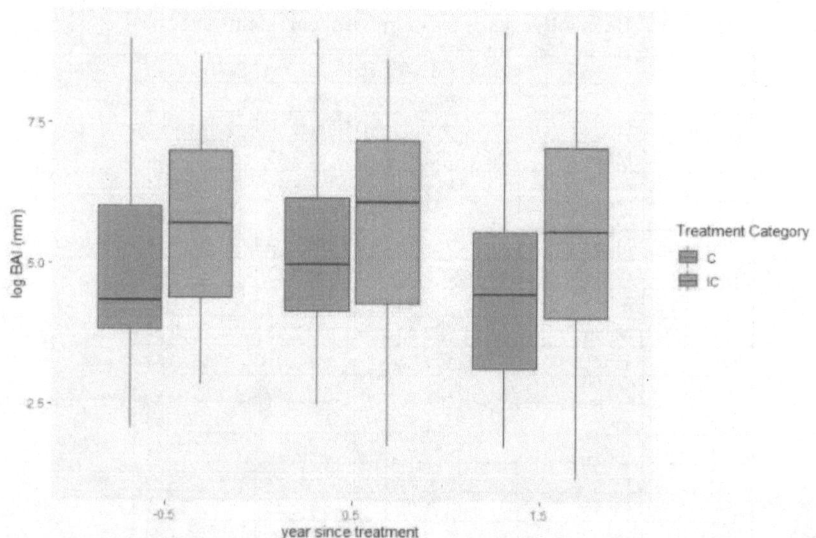

FIGURE 3.5 Basal area increments over time since treatment

TABLE 3.7 Descriptive statistics of growth increments over time

		2018-2019	2019-2020	2020-2021
		Time relative to cuts (time zero)		
Treatment	Stat	-0.5 years	+0.5 years	+1.5 years
C	Mean	752.73	870.45	854.64
BAI	sd	1899.05	1956.20	2116.94
	min	7.84	11.96	5.64
	max	7769.38	7911.09	9078.64
	range	7761.54	7899.13	9073.00
	n	29	29	29
IC	Mean	807.43	947.13	900.02
BAI	sd	1127.54	1260.91	1518.67
	min	17.34	0	3.22
	max	5876.32	5502.70	9120.64
	range	5858.97	5502.70	9117.42
	n	99	99	99
C	Mean	1.99	2.29	2.037
diameter	sd	2.41	2.46	2.64
	min	0.32	0.39	0.27
	max	9.95	10.04	10.75
	range	9.63	9.65	10.48
	n	29	29	29
IC	Mean	2.58	2.77	2.55
diameter	sd	1.92	2.11	2.24
	min	0.47	0	0.20
	max	8.65	8.37	10.78
	range	8.18	8.37	10.57
	n	99	99	99

Discussion

Initial Conditions

The initial tree size appeared to be very dependent on the species of the tree with tulip poplar being the largest, and elms being the smallest on average (Table 3.1 and 3.2). Larger diameter trees also had fewer overtopping trees that were either still living or targeted for girdling. The larger diameter trees show relationships with the canopy assessment variables (crown position, crown light exposure, and canopy position) that correspond to individuals with canopies that are taller, receive more light, and experience less competition than smaller diameter trees (Table 3.3, see Methods section in Chapter II for canopy health assessment variables).

The invasive shrub cover has no significant relation to the initial size, indicating that smaller trees do not have more invasive shrubs around them than larger trees. The overall tree condition is important to initial size of the target tree, with trees that are larger exhibiting better form and less damage than the smaller trees. Larger trees appear to also have soil with a greater moisture content and a more open canopy (Table 3.3). This would indicate that larger trees have had a history of greater access to water and light throughout their lifetime. These results provide a better characterization for the Working Woods area, especially considering the initial size and canopy position of the species in this area.

Pre-treatment Growth

Growth that occurred between the beginning of the growing season in 2018 and the beginning of the growing season of 2019 appears to depend on the initial size, the species and where the tree is located (Table 3.3 and Figure 3.2). This indicates that the growth occurring in Working Woods is highly heterogenous between individual trees, and that not all species or sizes of trees are growing at the same rate. For example, sugar maples are the only species that did not follow the general trend of larger sized trees growing more than smaller trees (Figure 3.2A). Additionally, target trees that had more live overtopping trees and overtopping trees that were targeted for girdling grew less than trees that did not have as much overtopping trees (Table 3.3). Ideally, this early trend would indicate that this treatment may help the trees that have more overtopping, competing neighbors. However, this may also require the girdled trees to die fully before an increased growth rate is observed. This could be because girdled trees are found to take several years to completely die and cease leafing out (Fassnacht & Steele, 2016), resulting in a lag in increased light availability. Additionally, another study which performed more intense harvesting methods found that the residual trees showed an increase in growth but only after 4 years after the treatment had been applied (Jones et al., 2009). This could mean that even once a girdled tree is dead, it could be several additional years until an increase in growth rate is observed.

Girdling Treatments Results Have Minimal Impacts on Growth

Our results show that girdling does not appear to have strong or consistent effects on the growth of residual trees within the first few years after treatment. Only elm trees in the understory showed some signs of responding positively to girdling (Fig 3.3). The lack of apparent girdling effects on growth, or interactions between girdling and factors like initial tree size or species identity, is perhaps not surprising after considering two other results from this study. First, many of the girdled trees were still living even 2.5 years after girdling. Second, the effects of girdling on canopy structure and light availability and competition appear to be minimal (Chapter II). Longer-term study in Working Woods is necessary to follow that impacts of girdling on tree growth, and to identify whether species with different traits or trees of different sizes and positions in the understory will ultimately have different growth responses to girdling.

As previous studies have shown, even with greater girdling than our study used, girdling did not result in a change in growth one season after treatment application

(Grigri et al., 2020). Even studies that have performed 'more intense' harvesting methods (e.g. selection cutting) find that many trees (including sugar maples) exhibit a three year or greater lag in growth response once released from competition (Jones et al., 2009; Jones & Thomas, 2004). Other studies have found that when 65% of the LAI of the surrounding canopy was girdled, a small increase in the wood primary production was observed after just one growing season post-girdling (Grigri et al., 2020). However, less disturbed areas in this study did not have a significant difference in the annual wood primary production compared to the more intensely disturbed areas and is attributed to the girdled trees continuing to fix carbon throughout the growing season. This result and the other study which observed lags in growth response after treatment would indicate that the Working Woods area requires continued observations. Fewer trees in Working Woods have been girdled compared to other girdling experiments, and as studies that performed 'more intense' treatments have shown, that the residual trees only show increases in growth several years after treatment.

Shrub Cover does not Have Strong Effects on Growth

The invasive shrub cover averaged over each growing period did not appear to have an effect on the growth, either pre or post-treatment. Thus, applying TSI improvements does not increase the annual growth of this population of trees within the first few years of shrub removal. Future analyses of sub-annual growth rates could be performed to identify if the presence of abundance of shrubs might have effects on growth during shorter time scales, such as during a heat wave or drought, when resource competition could be more impactful. Future analyses should also assess whether the impacts of shrub cover might depend on the size of the focal tree, as resource competition with shrubs might matter for smaller, understory trees, but not for larger, canopy dominant trees.

Land Manager Recommendations

Using girdling (and the occasional felling) of trees does not appear to be an effective method for promoting the growth of mid and understory trees, at least not within the first few years of treatment. The only species that showed a significant positive relation to the girdled intensity were elm trees that were composed of entirely understory individuals. Additionally, as other studies have shown, a greater intensity of girdling may be required for this response (Grigri et al., 2020).

However, this treatment should, given enough time, create snags which have been associated with various forest health metrics (Fassnacht & Steele, 2016; Sandström et al., 2019). Considering that management practices tend to result in increases to understory light (despite observations from Chapter II suggesting light condition changes are minorly influenced by girdling) it might be beneficial for continued TSI management strategies to prevent invasive shrub cover from increasing (Taylor et al., 2017). If the goal is to improve ecosystem services and the overall health of the forest then this girdling treatment, combined with invasive species management may be an effective solution.

In addition to ensuring that proper management techniques are used to reach the specific goals of the forested area, land managers should also be aware of how their land may respond to climate change. As temperatures increase, precipitation patterns will change in turn either resulting increased rainfall, or more intense drought events. The precipitation changes might result in less water available for current trees, resulting in additional stressors limiting tree growth as neighboring trees compete for soil moisture (Keenan, 2015). These secondary forests are likely to respond in unique ways based upon their ecosystem legacies, and the management applications used. Management applications will likely need to be adjusted to ensure that forests are resilient to climate change, and can still be used for timber production (Keenan, 2015). Especially as the tree species are likely to change as a result of management practices, climate change has the potential to cause large species shifts in current temperate secondary forests (Frelich et al., 2020). As managers adapt practices to these changing conditions, increasing species diversity has been shown to provide the greatest community resistance to climate change (Millar et al., 2007). It could prove beneficial to managers of secondary forests to increase the species diversity of these areas that tend to struggle with lower species diversity (Flinn & Marks, 2007).

www.ingramcontent.com/pod-product-compliance
Lightning Source LLC
LaVergne TN
LVHW041230150826
845673LV00008B/2334

* 9 7 8 3 3 8 4 2 8 1 4 8 7 *